Zebrafish: A Comprehensive Study

Zebrafish: A Comprehensive Study

Editor: Donald Clayton

STATES
ACADEMIC PRESS
www.statesacademicpress.com

States Academic Press,
109 South 5th Street,
Brooklyn, NY 11249, USA

Visit us on the World Wide Web at:
www.statesacademicpress.com

ISBN: 978-1-63989-575-5 (Hardback)

Cataloging-in-Publication Data

Zebrafish : a comprehensive study / edited by Donald Clayton.
   p. cm.
Includes bibliographical references and index.
ISBN 978-1-63989-575-5
1. Zebra danio. 2. Danio. 3. Zoology. I. Clayton, Donald.
QL638.C94 Z43 2022
597.482--dc23

Table of Contents

Preface

It is often said that books are a boon to mankind. They document every progress and pass on the knowledge from one generation to the other. They play a crucial role in our lives. Thus I was both excited and nervous while editing this book. I was pleased by the thought of being able to make a mark but I was also nervous to do it right because the future of students depends upon it. Hence, I took a few months to research further into the discipline, revise my knowledge and also explore some more aspects. Post this process, I begun with the editing of this book.

The zebrafish is a tropical freshwater fish belonging to the carp family of the order Cypriniformes. It is a popular aquarium fish native to South America. The name comes from the horizontal blue stripes on their body which resembles a zebra. The zebrafish is an important and widely used vertebrate model organism in scientific research, particularly drug development. It has remarkable regeneration abilities, and has been modified by scientists to produce many transgenic strains. The zebra fish shares 70% of its genes with the humans and has many organs similar to that of humans. Hence any type of disease which affects human body parts can also be remodelled in a zebrafish. This book provides comprehensive insights into the study of zebrafish. From theories to research to practical applications, case studies related to all contemporary topics of relevance to this field have been included herein. A number of latest researches have been included in this book to keep the readers up-to-date with the global concepts in this area of study.

I thank my publisher with all my heart for considering me worthy of this unparalleled opportunity and for showing unwavering faith in my skills. I would also like to thank the editorial team who worked closely with me at every step and contributed immensely towards the successful completion of this book. Last but not the least, I wish to thank my friends and colleagues for their support.

Editor

Excitation and Excitation-Contraction Coupling of the Zebrafish Heart: Implications for the Zebrafish Model in Drug Screening

Matti Vornanen, Jaakko Haverinen and
Minna Hassinen

Abstract

There are several similarities and differences in electrical excitability between zebrafish and human ventricles. Major ion currents generating ventricular action potentials are largely the same in human and zebrafish hearts with some exceptions. A large T-type calcium current is unique to the zebrafish ventricle (absent in human ventricle), and two potassium currents (I_{Ks} and I_{to}) may be absent in zebrafish ventricular myocytes. However, there are substantial differences among alpha subunit isoforms of the ion channel families or subfamilies (e.g. zebrafish Kv11.2 vs. human Kv11.1; zebrafish Kir2.4 vs. human Kir2.1) between human and zebrafish hearts. Contraction of zebrafish ventricle is strongly dependent on extracellular calcium, while human ventricle relies heavily on calcium stores of the sarcoplasmic reticulum. These differences may affect the use of zebrafish as a model in drug screening and safety pharmacology.

Keywords: cardiac action potential, ion currents, ion channels, drug screening, e-c coupling

1. Introduction

Zebrafish (*Danio rerio*), a tropical freshwater fish species, is a popular vertebrate model and widely used to resolve diverse research questions in developmental biology and genetics, human diseases, environmental toxicology and several other disciplines. The advantages of the zebrafish model are research technical (e.g. well-annotated and easily modifiable genome, transparency of embryos), economical (cost and ease of maintenance, large number of offspring and short generation time) and ethical (replacement of mammalian models—3R principle) [1, 2]. Those qualifications have made zebrafish an interesting object in studies, where new molecules are searched and selected for drug development programs [3–5]. Potentially zebrafish could be a high throughput and relatively inexpensive *in vivo* model for screening therapeutically effective and nontoxic candidate molecules for drug development programs. Indeed, great expectations are set on the zebrafish model, which is sometimes regarded as an ideal system for preclinical screening of cardiovascular drugs [6]. The expectations are based on the conserved properties of cardiac physiology between humans and zebrafish, such as the similarities in the shape of ventricular action potential (AP) and heart rate [7–9]. The documented responses of zebrafish hearts (e.g. bradycardia, atrioventricular block, prolongation QT interval of electrocardiogram) to the inhibitors of human *ether-à-go-go*-related (KCNH2) channel provide some credence to those expectations, even though sensitivity and specificity of the responses are not optimal [9–13]. In the screening of cardiovascular drugs, the *in vivo* zebrafish model has the advantage that all cardiac ion channels are simultaneously exposed to the compound, thereby allowing phenotype-based screening. However, in order to provide an accurate mode for the human heart, molecular composition, voltage-dependence and gating kinetics of ion channels of the zebrafish heart should closely match those of the human heart. Unfortunately, the ionic and molecular bases of electrical excitability of the zebrafish heart are still unsatisfactorily known. This is a significant shortcoming, since the requirements set for effective and safe drugs are extremely rigorous, and safety evaluation necessitates exact knowledge about the mode of drug action [14]. Those requirements are delineated in Comprehensive in vitro Proarrhythmia Assay (CiPA) initiative for cardiac safety evaluation of new drugs, which provides an accurate mechanistic-based assessment of proarrhythmic potential [14–16]. Rational evaluation of drug toxicity in the zebrafish model is not possible before ion currents and channels of the zebrafish cardiac myocytes are known in sufficient detail. The present overview compares ion current and ion channel compositions of zebrafish and human ventricles in order to indicate similarities and differences between the fish model and the human heart, and gaps in our knowledge of the zebrafish cardiac excitability and excitation-contraction (e-c) coupling. These issues have also been discussed in other recent reviews [17–19].

2. Cardiac action potential

Contraction of cardiac myocytes is triggered by electrical excitation of myocyte sarcolemma in the form of cardiac AP. Propagation of AP through the heart can be recorded as electrocardiogram (**Figure 1A**). Each functionally different cardiac tissue has a characteristic AP shape generated by the tissue-specific ion currents and ion channel compositions. The five different phases of the mammalian ventricular AP—with the exception of phase 1 fast repolarization—are readily discernible in the zebrafish ventricular AP (**Figure 1B**). Similar to the human ventricular AP (but unlike the murine AP), the zebrafish ventricular AP has a distinct plateau phase (phase 2) at positive voltages [8, 9, 20] (**Figure 1B**). Indeed, the only major difference between zebrafish and human ventricular APs is the absence of the fast phase 1 repolarization

A

B

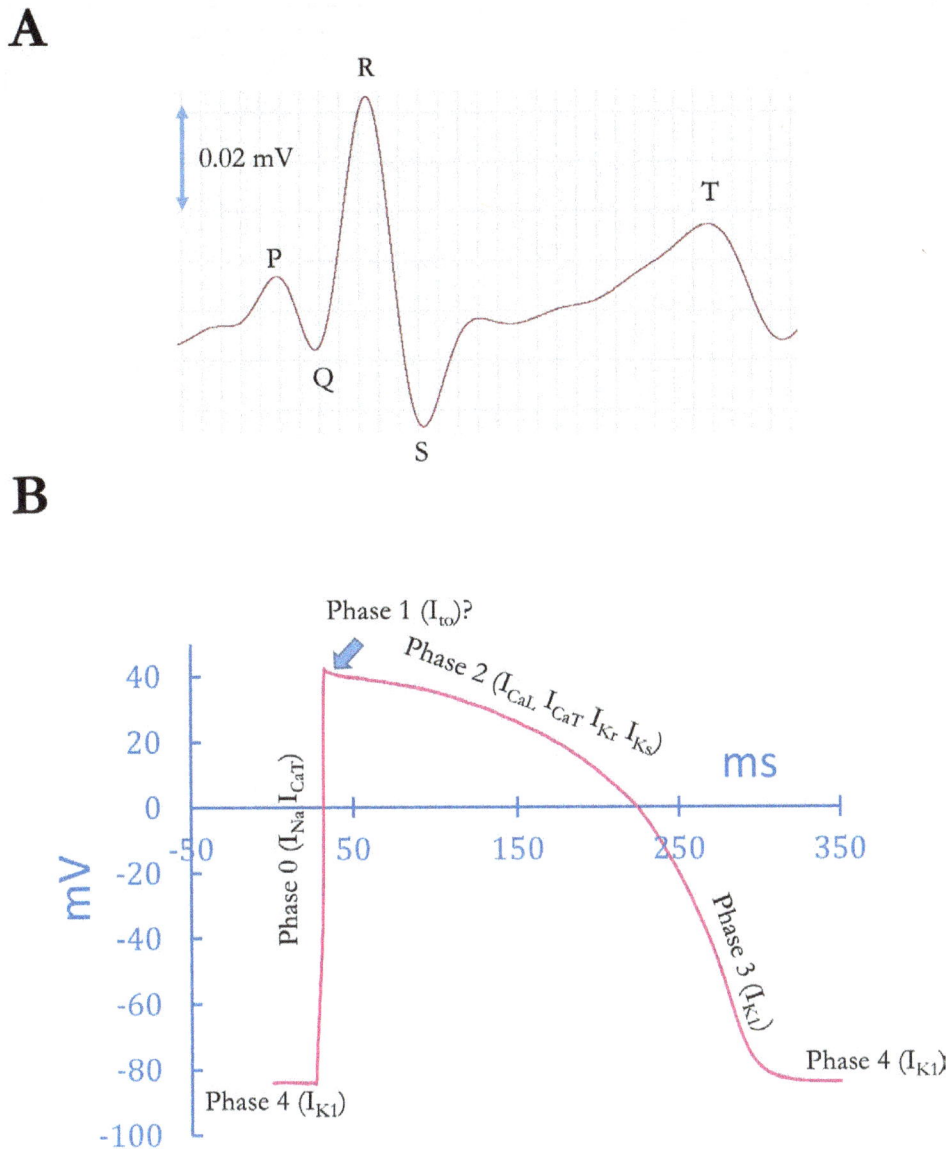

Figure 1. Electrocardiogram (A) and ventricular action potential (B) of the zebrafish heart at 26°C. Electrocardiogram was recorded from surface of spontaneously beating heart *in vitro*. Ventricular action potential was recorded from an enzymatically isolated cardiac myocyte with patch-clamp technique. The main ion currents responsible for different phases of ventricular action potential are also shown. I_{Na}, Na$^+$ current; I_{CaL}, L-type Ca^{2+} current; I_{CaT}, T-type Ca^{2+} current; I_{Kr}, the fast component of delayed rectifier K$^+$ current; I_{Ks}, the slow component of the delayed rectifier K$^+$ current; I_{K1}, the inward rectifier K$^+$ current.

in the zebrafish AP [19]. This may be due to the absence of the transient outward K$^+$ current (I_{to}) in zebrafish ventricular myocytes [8].

Zebrafish are ectothermic vertebrates and therefore their AP characteristics may change depending on the rearing temperature of the fish, as has been reported for several teleost fish species [21, 22]. In the adult zebrafish, reared at 28°C, the duration of ventricular AP (APD$_{50}$; at 28°C) is about 30% shorter than that of the human ventricular AP at 37°C (**Table 1**). At 36°C, the duration of zebrafish ventricular AP is only 80 ms, i.e. about 25% of the duration of human ventricular AP at 37°C. The shorter AP of the zebrafish heart may be associated with

	Zebrafish	Reference	Human	Reference
Relative heart size (% of body mass)	0.1	[48]	0.64	[50]
Diastolic/systolic blood pressure (mm Hg)	0.42/2.51	[70]	70/125	[71]
Myocyte size (ventricle) (pF)	26–33	[8, 39]	117–227	[72, 73]
T-tubuli	No	[20, 48]	Yes	[67, 74]
Role of CICR[1] in e-c coupling (%)	15	[48]	~70	[75, 76]
Myofibril location	Subsarcolemmal	[47]	Throughout the myocyte	
Ventricular AP duration (ms)	~240 at 28°C ~80 at 37°C	[19]	~330 at 37°C	[77]
Resting heart rate (bpm)	120–130 at 28°C ~287 at 37°C	[8, 19, 23]	60–80 bpm at 37°C	[71]

[1]CICR, Ca^{2+}-induced Ca^{2+} release.

Table 1. Basic characteristics of the zebrafish heart and cardiac myocytes in comparison to those of the human heart.

the higher heart rate of the fish, which at 28°C is about double (120–130 beats per minute) and at 37°C about quadruple (287 beats per minute) the human resting heart rate [19, 23] (**Table 1**). Temperature is an important environmental factor for an ectothermic vertebrate, which modifies cardiac gene expression and ion channel function [24–26]. Therefore, rearing and experimental temperatures should be carefully controlled and reported in zebrafish studies.

3. Ion currents and ion channels

Density and kinetics of ion currents must be such that chamber-specific APs are generated and can be adjusted to heart rates according to the circulatory demands. This is reflected in the composition of ion channel assemblies and their abundances in different cardiac chambers and in a species-specific manner [21]. This overview is limited to ventricular myocytes, since atrial ion currents/channels of the zebrafish heart are still relatively poorly known.

3.1. Sodium currents and channels

Sodium influx through the voltage-gated Na^+ channels initiates the all-or-none action potential (AP) of atrial and ventricular myocytes, when the current flow from the upstream cell depolarizes the membrane of the downstream cell to the threshold level (about −55 mV). At the threshold voltage, the density of inward Na^+ current (I_{Na}) exceeds the total density of the outward K^+ currents (I_K). The rapid opening of Na^+ channels generates a fast upstroke (depolarization) of AP and an overshoot to the level of about +40 mV (phase 0) [19, 20] (**Figure 1B**). Then I_{Na} quickly inactivates due to the closing of the inactivation gate of the channel.

The density of I_{Na} is the main determinant for the rate of AP propagation over the heart. The rate of AP upstroke in zebrafish ventricular myocytes at 28°C is about 130 V s^{-1} (RMP ~ −84 mV) (Haverinen et al., submitted), which is less than half of the rate of AP upstroke in human ventricular myocytes at 37°C [27]. These findings suggest that the density of ventricular I_{Na} is lower and the rate of AP propagation slower than in human ventricles at the species-specific temperatures (28°C vs. 37°C). However, a thorough analysis of the zebrafish I_{Na} is needed to reveal to what extent these differences are due to RMP (availability of Na$^+$ channel for opening), Na$^+$ channel density and kinetic properties of the cardiac Na$^+$ channels.

The zebrafish heart expresses eight different Na$^+$ channel alpha subunits. The main isoforms are Na$_v$1.5Lb (83.1% of the transcripts) and Na$_v$1.4b (16.2%), which are orthologues to the human cardiac Na$_v$1.5 (71.1% in the right ventricle) and skeletal Na$_v$1.4 channels, respectively [28] (**Table 2**). Na$_v$2.1 is abundantly expressed in the human right ventricle (27.8%), but seems to be absent in zebrafish ventricular myocytes. Unlike the mammalian cardiac I_{Na}, which is tetrodotoxin-resistant (IC$_{50}$ about 1 μM), the zebrafish I_{Na} is more than two orders of magnitude more sensitive to tetrodotoxin (IC$_{50}$ about 6 nM) (Haverinen et al., submitted), similar to the I_{Na} of other fish species [29, 30]. Thus, there is a remarkable difference in tetrodotoxin-sensitivity and some minor differences in Na$^+$ channel composition between zebrafish and human hearts.

3.2. Calcium currents and channels

The vertebrate heart usually has two main types of Ca^{2+} currents, a high-threshold or long-lasting L-type current (I_{CaL}) and a low-threshold or transient T-type (I_{CaT}) current. The former is activated at voltages more positive than −40 mV and with the peak amplitude at about +10 mV, while the latter is generated already at −60 mV and with the peak current amplitude at about −30 mV [31, 32].

I_{CaL} is the main I_{Ca} of atrial and ventricular myocytes. It has a significant physiological function in maintaining the long plateau (phase 2) of the cardiac AP and mediating Ca^{2+} influx into the myocyte (**Figure 1B**). I_{CaL} contributes to the Ca^{2+} transient, which sets cardiac contraction in motion, either directly by increasing cytosolic Ca^{2+} concentration or triggering a further release of Ca^{2+} from the sarcoplasmic reticulum (SR) (for more details see excitation-contraction coupling). The mean density of I_{CaL} in ventricular myocytes of the zebrafish heart at 28°C is 6–8 pA pF^{-1}, which is about double the density of the human ventricular I_{CaL} at 35°C (3–4 pA pF^{-1}) [8, 20, 33, 34]. This difference may signify a larger role of sarcolemmal Ca^{2+} influx in e-c coupling of the zebrafish heart [33]. In the human ventricle, the main L-type Ca^{2+} channel isoform is Ca$_v$1.2, which represents 98.4% of the total Ca^{2+} channel transcripts in the right ventricle [35] (**Table 2**). In the zebrafish ventricle, seven different L-type Ca^{2+} alpha subunits are expressed, among them three paralogue pairs [36]. Similar to the human ventricle, Ca$_v$1.2 is the most abundant L-type Ca^{2+} channel isoform in zebrafish ventricle consisting of 38.3% of the all Ca^{2+} channels transcripts.

In vertebrate hearts I_{CaT} is a sizeable current in nodal myocytes and it may be also present in atrial myocytes, but it is usually absent in ventricular myocytes. In this respect, the zebrafish is clearly different. A characteristic feature for zebrafish ventricular myocytes is a large I_{CaT} with a

Ion current	Ion channels[*]	Ion channels[**]
	Zebrafish	Human
I_{Na}	**$Na_v1.5Lb$ (83.1%)**	**$Na_v1.5$ (71.1%)**
	$Na_v1.4b$ (16.2%)	$Na_v2.1$ (27.8%)
	$Na_v1.1b$ (0.48%)	$Na_v1.3$ (7.1%)
	$Na_v1.6b$ (0.12%)	$Na_v1.1$ (2.7%)
	$Na_v1.6a$ (0.03%)	$Na_v1.7$ (1.4%)
	$Na_v1.1a$ (0.02%)	
	$Na_v1.4a$ (0.01%)	
	$Na_v1.5La$ (0.01%)	
I_{Ca}		
I_{CaL}	**$Ca_v1.2$ (38.3%)**	**$Ca_v1.2$ (98.4%)**
	$Ca_v1.3a$ (0.07%)	$Ca_v1.3$ (0.039%)
	$Ca_v1.1a$ (2.63%)	
I_{CaT}	**$Ca_v3.1$ (54.8%)**	$Ca_v3.1$ (0.14%)
	$Ca_v3.2a$ (0.06%)	**$Ca_v3.2$ (1.45%)**
	$Ca_v3.2b$ (0.03%)	
$I_{CaP/Q}$	$Ca_v2.1b$ (3.84%)	
I_K (voltage-gated)		
I_{Kr}	$K_v11.1a$ (0.1%)	**$K_v11.1$[2] (54.3%)**
	$K_v11.1b$ (0.1%)	
	$K_v11.2a$[1] (84.6%)	
	$K_v11.2b$[1] (0.3%)	
	$K_v11.3$ (0.2%)	
I_{to}		$K_v1.5$ (26.1%)
I_{Ks}	$K_v7.1$ (14.6%)	$K_v7.1$ (15.7%)
I_{Kur}	Not examined	**$K_v4.3$ (12.1%)**
I_{K1} (inward rectifying)		
	Kir2.1a (0.6%)	**Kir2.1 (46.5%)**
	Kir2.1b (0.005%)	
	Kir2.2a (6.3%)	Kir2.2 (28.9%)
	Kir2.2b (0.1%)	
	Kir2.3 (0.04%)	Kir2.3 (24.5%)
	Kir2.4 (93.0%)	

[1]KCNH6 (Zebrafish erg).
[2]KCNH2 (Human erg).
[*]Zebrafish results are from [19, 36] and unpublished results of Hassinen et al.
[**]Human results are from [35].

Table 2. Major ion currents and ion channel transcripts of zebrafish ventricle in comparison to those of the human ventricle.

current density almost equal to that of I_{CaL} [8, 37]. T-type Ca^{2+} channels pass Ca^{2+} influx at more negative voltages than L-type Ca^{2+} channels. Therefore, they may contribute to upstroke and early plateau of the ventricular AP. Although I_{CaT} inactivates faster than I_{CaL}, it allows a significant sarcolemmal Ca^{2+} entry into zebrafish ventricular myocytes. During a 300 ms depolarizing pulse to −30 mV the Ca^{2+} influx through T-type Ca^{2+} channels is about 35% of the Ca^{2+} influx of L-type Ca^{2+} channels during 300 ms pulse to +10 mV [36]. Therefore, I_{CaT} may have a significant role in e-c coupling of zebrafish ventricular myocytes. T-type Ca^{2+} channels are abundantly expressed in the zebrafish ventricle constituting majority of the transcripts (about 55%) of the total Ca^{2+} channel population (**Table 2**). Altogether five alpha subunits of the T-type (Ca_v3) family are expressed in the zebrafish ventricle. $Ca_v3.1$ (alpha1G) is clearly the dominant isoform, not only among T-type Ca^{2+} channels, but also among all the cardiac Ca^{2+} channel types with the transcript abundance of 54.8%. The other T-type alpha subunits are expressed only in trace amounts. In human ventricular myocytes, the T-type Ca^{2+} channels are very weakly expressed. $Ca_v3.1$ and $Ca_v3.2$ alpha subunits constitute together less than 2% of the Ca^{2+} channels transcripts in the human right ventricle [35] (**Table 2**). Taken together the prominent expression of I_{CaT} in zebrafish ventricular myocytes is one of the most striking differences in ion channel composition between zebrafish and human hearts. The exact role of I_{CaT} in excitation and e-c coupling of zebrafish ventricular myocytes needs to be examined in detail.

Overall, the diversity of Ca^{2+} channels in zebrafish ventricle is larger than in the human ventricle. Most notably, T-type Ca^{2+} channels are more abundant than L-type Ca^{2+} channels.

3.3. Potassium currents and channels

Outward potassium currents (I_K) are repolarizing, i.e. they maintain negative resting membrane potential (RMP) and limit the duration of cardiac AP.

3.3.1. Inward rectifier K^+ currents and channels

The inward rectifier K^+ current (I_{K1}) maintains RMP of atrial and ventricular myocytes and provides K^+ efflux for the final phase 3 repolarization of AP [38] (**Figure 1B**). This current is generated by the Kir2 subfamily channels in vertebrate hearts. Characteristic for the inward rectifier K^+ channels of the Kir2 family is that they pass outward I_{K1} in the voltage range from about −80 to 0 mV with the peak current at −59 mV in zebrafish ventricular myocytes [39]. At RMP, the net K^+ flux at the RMP is almost zero, but the outward I_{K1} activates instantaneously on depolarization of sarcolemma and generates a fast outward surge of I_{K1} at the rising phase of AP. At the plateau voltage, K^+ efflux is small due to the voltage-dependent block of the channel pore by intracellular polyamines and Mg^{2+} ions. When membrane potential starts to repolarize (due to the activation of I_{Kr} and I_{Ks} and inactivation of I_{Ca}), the polyamine block of Kir2 channels relaxes and K^+ efflux through Kir2 channel accelerates repolarization at phase 3. Different members of the Kir2 family differ in the ease with which they allow K^+ efflux through the sarcolemma. Large difference in Ba^{2+}-sensitivity between Kir2 isoforms suggests that their interaction with potential medicinal drug molecules might also differ [39].

The main Kir2 channel isoform of the zebrafish ventricle is Kir2.4, as in the most teleost fishes studied thus far [39]. In the zebrafish atrium Kir2.2a channels are abundantly expressed in addition to Kir2.4 channels. There are striking differences in isoform composition between zebrafish and human ventricles. The main isoform in the human ventricles is Kir2.1 (46.5% of the transcripts in the right ventricle), which appears only in trace amounts (0.8%) in the zebrafish ventricle. Kir2.3 is abundantly expressed in human ventricle (24.5%), but almost totally absent in the zebrafish ventricle (**Table 2**). Since the Kir2 isoforms differ in their rectification properties those differences in isoform composition are likely to have functional consequences for repolarization of AP.

3.3.2. Voltage-gated K^+ currents and channels

Voltage-gated K^+ currents provide sarcolemmal K^+ efflux for repolarization of AP. Several different voltage-gated K^+ currents are expressed in human ventricle including fast and slow components of the delayed rectifier current (I_{Kr} and I_{Ks}, respectively) and transient outward current (I_{to}) [27, 40, 41]. Transcripts for the ultra-rapid component of the delayed rectifier channel ($K_v1.5$) are expressed in the human ventricle, but the current seems to be specific for atrial myocytes and not expressed in the ventricles [42]. Much less is known about the voltage-gated K^+ currents of the zebrafish heart. Similar to the human ventricle, I_{Kr} seems to be the main repolarizing current also in zebrafish ventricle [8, 37, 43]. Until now, I_{Ks} has not been recorded in zebrafish cardiac myocytes, even though transcripts of the $K_v7.1$ (KCNQ1) channel are expressed in the zebrafish ventricle (Hassinen et al., unpublished) (**Table 2**). Neither has been I_{to} found in zebrafish cardiac myocytes [8]. These findings suggest that repolarization of zebrafish ventricular myocytes might be more strongly dependent on single voltage-gated K^+ current, I_{Kr}, than human ventricular myocytes. However, a closer examination of the K^+ ion composition of zebrafish ventricular myocytes is needed to verify/falsify these assumptions.

I_{Kr} is the dominant repolarizing current in human and zebrafish ventricular myocytes. Notably the current is generated by different isoforms in human and zebrafish heart. In the human cardiac myocytes, I_{Kr} flows through the erg1 (Herg or KCNH2 or $K_v11.1$) channels (**Table 2**). In zebrafish myocytes I_{Kr} is generated almost exclusively by KCNH6 ($K_v11.2a,b$ or zebrafish erg) channels [19, 44], although it is often referred to as an orthologue to human KCNH2 channels [9, 11, 45]. Indeed, the expression of $K_v11.1a,b$ transcripts in the zebrafish ventricle is only 0.2%, while $K_v11.2a,b$ channel transcripts constitute 84.9% of the total voltage-gated K^+ channel alpha subunit population. Transcripts of the $K_v7.1$ channels form about 15% of the total transcripts of the voltage-gated K^+ channels in both human and zebrafish ventricles (**Table 2**). $K_v1.5$ channels are expressed in the human ventricle, but no reports exist about any zebrafish orthologues.

4. Excitation-contraction coupling

There are prominent differences in size, shape and fine structure between zebrafish and human ventricular myocytes (**Table 1**). Zebrafish ventricular myocytes are 5–10 times smaller (26.1–33.3 pF) than human ventricular myocytes (117–227 pF). Zebrafish ventricular myocytes are almost as long as human ventricular myocytes, but are much thinner [20, 46]. In human ventricular myocytes, myofibrils are evenly distributed throughout the myocyte, while in zebrafish ventricular myocytes myofibrils locate immediately under the sarcolemma [47]. Due to the small diameter of ventricular myocytes and cortical location of myofibrils T-tubuli are unnecessary for cellular signaling and probably therefore completely absent in zebrafish myocytes [20, 48].

The marked differences in myocyte size and structure appear as significant differences in the excitation-contraction (e-c) coupling between zebrafish and human (mammalian) cardiac myocytes (**Figure 2**). The contraction of human ventricular myocyte mainly relies on intracellular Ca^{2+} stores of the sarcoplasmic reticulum (SR) in generating cytosolic Ca^{2+} transients. A small Ca^{2+} influx through L-type Ca^{2+} channels triggers a large Ca^{2+} release via ryanodine-sensitive Ca^{2+} release channels of the SR so that about 77% of the cytosolic Ca^{2+} transient originates from the SR [49, 50]. In ventricular myocytes of the zebrafish heart Ca^{2+} release from the SR makes only about 15% of the total Ca^{2+} transient [48]. Voltage-dependence of cell shortening and Ca^{2+} transients also suggest that sarcolemma Ca^{2+} influx is the main source of cytosolic Ca^{2+} in zebrafish cardiomyocytes [51]. In human cardiac myocytes, voltage dependence of the Ca^{2+} transients is bell-shaped reflecting the voltage dependence of the trigger for Ca^{2+} release from the SR, I_{CaL} [52]. In zebrafish myocytes both I_{CaL} and Na^+-Ca^{2+}-exchange directly contribute to

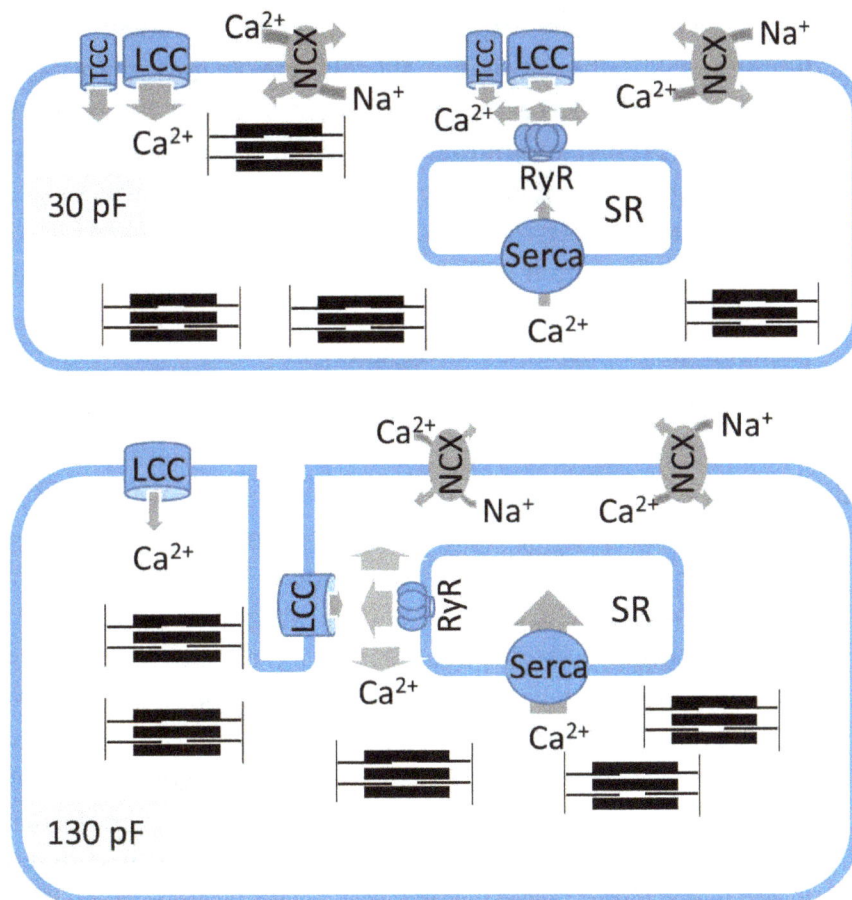

Figure 2. A simplified scheme about excitation-contraction coupling of the zebrafish ventricle (top) in comparison to that of the human ventricle (bottom). Main influx and efflux pathways of Ca^{2+} during contraction and relaxation of the ventricular myocyte. LCC, L-type Ca^{2+} channel; TCC, T-type Ca^{2+} channel; NCX, Na^+-Ca^{2+} exchange; RyR, ryanodine receptor; SR, sarcoplasmic reticulum; Serca, sarco-endoplasmic reticulum Ca^{2+}-ATPase.

cytosolic Ca^{2+} transient resulting in monophasic voltage dependence of cell shortening and Ca^{2+} transients [51]. Sarcolemmal Ca^{2+} influx via I_{CaL} and I_{CaT} during a single twitch is almost 130 μM L^{-1} from which about 32% occurs through T-type Ca^{2+} channels [36]. These differences in Ca^{2+} handling are associated with 71% lower expression of ryanodine receptors in the zebrafish ventricle in comparison to mammalian (rabbit) ventricle, whereas little differences exist in the SR Ca^{2+} content [48]. Ca^{2+} sensitivity of ryanodine receptors of the fish heart is often low in comparison to that of mammalian cardiac ryanodine receptors [53], which might also contribute to the small SR Ca^{2+} release in zebrafish cardiomyocytes [48]. Overall, the main differences in e-c coupling between zebrafish and human ventricular myocytes are the smaller role of intracellular Ca^{2+} stores of the SR, the presence of large I_{CaT} and the absence of T-tubuli in the zebrafish myocytes.

5. Implications for the use of zebrafish in drug screening

The use of animal models for studies of human cardiac electrophysiology is based on the similarity of animal and human hearts concerning ion current densities, ion channel compositions and mechanisms of ion channel regulation by rate changes and autonomic nervous system agonists [54, 55]. However, electrophysiological properties of cardiac myocytes are species-specific and significantly different even between mammalian species (e.g. human vs. dog, rabbit and guinea pig) [54, 55]. Therefore, it is necessary to consider, whether the noticed differences in electrophysiology between zebrafish and human ventricular myocyte might affect the status of zebrafish as a model for drug screening and safety pharmacology. In this respect, quantitative differences in repolarizing currents between model and human hearts are regarded significant [54, 55]. Since ion channel isoforms often differ concerning activation and inactivation kinetics, voltage-dependence and drug affinity [39, 43, 56–58], isoform-composition of the expressed ion channels is also expected to be important for drug screening.

The human erg (KCNH2) channel is known for its propensity of being blocked by wide variety of small molecules, which may lead to AP prolongation and lethal cardiac arrhythmias. Therefore, preclinical drug screening procedures aim to assess drug-induced inhibition of I_{Kr} and prolongation of AP and QTc interval of electrocardiogram in order to remove proarrhythmic molecules from the drug development programs. However, drugs which inhibit I_{Kr} do not always produce QT prolongation due to simultaneous inhibition of the depolarizing I_{Na} and I_{Ca} currents, or since other ion channels provide a repolarization reserve to compensate for I_{Kr} inhibition. Proarrhythmic effects appear only when the drug changes the balance between inward and outward currents. To improve preclinical drug screening, the consortium of international stakeholders has recently launched an initiative called Comprehensive in-vitro Proarrhythmia Assay (CiPA). One of the central tenets of this initiative is that drug molecules are tested *in vitro* against multiple ion channels. Indeed, recent studies have indicated that the assessment of drug affinity toward multiple ionic targets improves the prediction of proarrhythmia risk in comparison to the sole I_{Kr} analysis [59, 60]. For example, if the outward I_{Kr} and inward I_{CaL} are inhibited with a similar IC_{50} value (e.g. in the case of verapamil), the proarrhythmia risk is low.

Potential confounding factors in the applicability of zebrafish as a model is the presence of large I_{CaT} and the putative absence of I_{Ks} and I_{to}. Since different channel isoforms of the same subfamily may have different electrophysiological properties and drug affinities, the erg (human KCNH2 vs. zebrafish KCNH6) and Kir2 (human Kir2.1 vs. zebrafish Kir2.4) channel isoform compositions may be also important. Zebrafish might be a useful high-throughput drug screening platform with the advantages of both phenotype screening, if it fulfils the qualifications of the CiPA procedure. Therefore, the calibration routine of the CiPA initiative should be conducted on zebrafish cardiac myocytes [61]. This routine involves key cardiac ion channels (I_{Kr}, I_{CaL}, I_{Na}, I_{to}, I_{Ks} and I_{K1}), which should be examined under standardized voltage-clamp conditions for inhibition potency (IC_{50}) of 12 selected drugs of the "minimally acceptable" dataset [61]. These compounds are categorized into high, intermediate and low risk of torsades de pointes arrhythmia according to their currently known properties. The list of target channels should also include I_{CaT} in the zebrafish.

One more factor that may be important in regard to proarrhythmia potency of zebrafish heart is the e-c coupling of ventricular myocytes. Factors recognized as significant causes of cardiac arrhythmias in mammals include APs that are too long or too short. If APs are abnormally long, early afterdepolarizations during the AP plateau (Phase 2 or 3) may be provoked by reactivation of Ca^{2+} or Na^+ currents in the voltage "window," where all Ca^{2+} and Na^+ channels have not yet been inactivated and can be reactivated. In addition, early afterdepolarizations are promoted by spontaneous Ca^{2+} release from the SR that activates inward current via the reverse mode operation of Na^+-Ca^{2+} exchange [62]. APs that are too short can predispose the heart to delayed afterdepolarizations, which occur in early diastole (Phase 4), when spontaneous Ca^{2+} releases from the SR activate the inward Na^+-Ca^{2+}-exchange current. Afterdepolarizations may depolarize membrane potential to the AP threshold and induce extra systoles (triggered activity). Generation of delayed afterdepolarizations requires spontaneous Ca^{2+} release from the SR, which subsequently activates Ca^{2+} efflux via Na^+-Ca^{2+} exchange and membrane depolarization to the AP threshold [63]. In zebrafish heart, SR makes only a minor contribution to cardiac e-c coupling, possibly due to the low Ca^{2+} sensitivity of the ryanodine receptors. The relative independence of fish heart contraction from Ca^{2+}-induced Ca^{2+} release is expected to make the zebrafish heart relatively resistant against early and delayed afterdepolarizations and therefore less suitable as an arrhythmia model [18].

The similarities between human and zebrafish cardiac electrophysiology are often emphasized, while the differences are overlooked or neglected. However, it may be that those physiological functions and electrophysiological properties that are unique to the zebrafish heart are the most useful features for cardiac research. Perhaps the most spectacular example is the exceptional regeneration power of the zebrafish heart, which may reveal to us the molecular underpinnings needed to heal the damaged human heart [64]. Similarly, the exceptionally strong expression of the I_{CaT} in the zebrafish ventricle may provide a test system to examine the role of T-type Ca^{2+} channels in e-c coupling and its significance as a drug target. In the human heart, I_{CaT} is re-expressed and T-tubuli are lost, when the heart is subjected to pathological stressors that induce hypertrophy and failure [65–67]. Thus, zebrafish heart could be a "natural" model for testing proarrhythmic propensity of the drugs in the diseased heart. Structural (small cell size, absence of T-tubuli) and functional characteristics (minor role of SR in e-c coupling, presence of I_{CaT}) of the zebrafish ventricle are more like those of fetal or neonatal mammalian heart than those of the adult human heart [68, 69]. Therefore, the zebrafish heart might be a better model for fetal/neonatal and diseased human heart than for the adult human heart in drug screening.

Acknowledgements

This study was supported by a grant from Jane and Aatos Erkko Foundation to MV (project N0. 64579).

Conflict of interest

The authors declare that they have no conflict of interest.

Author details

Matti Vornanen*, Jaakko Haverinen and Minna Hassinen

*Address all correspondence to: matti.vornanen@uef.fi

Department of Environmental and Biological Sciences, University of Eastern Finland, Finland

References

[1] Kalueff AV, Echevarria DJ, Stewart AM. Gaining translational momentum: More zebrafish models for neuroscience. Progress in Neuro-Psychopharmacology & Biological Psychiatry. 2014;**55**:1-6

[2] Briggs JP. The zebrafish: A new model organism for integrative physiology. The American Journal of Physiology. 2002;**282**:R3-R9

[3] Chakraborty C, Hsu CH, Wen ZH, Lin CS, Agoramoorthy G. Zebrafish: A complete animal model for in vivo drug discovery and development. Current Drug Metabolism. 2009;**10**(2):116-124

[4] Parng C, Seng WL, Semino C, McGrath P. Zebrafish: A preclinical model for drug screening. Assay and Drug Development Technologies. 2002;**1**(1):41-48

[5] MacRae CA, Peterson RT. Zebrafish as tools for drug discovery. Nature Reviews Drug Discovery. 2015;**14**(10):721-731

[6] Qi M, Chen Y. Zebrafish as a model for cardiac development and diseases. Human Genetics & Embryology. 2015;**2015**

[7] Leong IUS, Skinner JR, Shelling AN, Love DR. Zebrafish as a model for long QT syndrome: The evidence and the means of manipulating zebrafish gene expression. Acta Physiologica (Oxford, England). 2010 Jul 1;**199**(3):257-276

[8] Nemtsas P, Wettwer E, Christ T, Weidinger G, Ravens U. Adult zebrafish heart as a model for human heart? An electrophysiological study. Journal of Molecular and Cellular Cardiology. 2010 Jan;**48**(1):161-171

[9] Arnaout R, Ferrer T, Huisken J, Spitzer K, Stainier DY, Tristani-Firouzi M, et al. Zebrafish model for human long QT syndrome. Proceedings of the National Academy of Sciences of the United States of America. 2007 Jul 3;**104**(27):11316-11321

[10] Langheinrich U, Vacun G, Wagner T. Zebrafish embryos express an orthologue of HERG and are sensitive toward a range of QT-prolonging drugs inducing severe arrhythmia. Toxicology and Applied Pharmacology. 2003;**193**(3):370-382

[11] Milan DJ, Peterson TA, Ruskin JN, Peterson RT, MacRae CA. Drugs that induce repolarization abnormalities cause bradycardia in zebrafish. Circulation. 2003;**107**(10):1355-1358

[12] Mittelstadt SW, Hemenway CL, Craig MP, Hove JR. Evaluation of zebrafish embryos as a model for assessing inhibition of hERG. Journal of Pharmacological and Toxicological Methods. 2008 Mar-Apr;**57**(2):100-105

[13] Tsai CT, Wu CK, Chiang FT, Tseng CD, Lee JK, Yu CC, et al. In-vitro recording of adult zebrafish heart electrocardiogram—A platform for pharmacological testing. Clinica Chimica Acta. 2011;**412**(21-22):1963-1967

[14] Cavero I, Guillon JM, Ballet V, Clements M, Gerbeau JF, Holzgrefe H. Comprehensive in vitro Proarrhythmia Assay (CiPA): Pending issues for successful validation and implementation. Journal of Pharmacological and Toxicological Methods. 2016;**81**:21-36

[15] Vicente J, Stockbridge N, Strauss DG. Evolving regulatory paradigm for proarrhythmic risk assessment for new drugs. Journal of Electrocardiology. 2016;**49**(6):837-842

[16] Crumb WJ, Vicente J, Johannesen L, Strauss DG. An evaluation of 30 clinical drugs against the comprehensive in vitro proarrhythmia assay (CiPA) proposed ion channel panel. Journal of Pharmacological and Toxicological Methods. 2016;**81**:251-262

[17] Genge CE, Lin E, Lee L, Sheng XY, Rayani K, Gunawan M, et al. The zebrafish heart as a model of mammalian cardiac function. Reviews of Physiology, Biochemistry and Pharmacology. 2016;**171**:99-136

[18] Verkerk AO, Remme CA. Zebrafish: A novel research tool for cardiac (patho)electrophysiology and ion channel disorders. Frontiers in Physiology. 2012 Jul 10;**3**:255

[19] Vornanen M, Hassinen M. Zebrafish heart as a model for human cardiac electrophysiology. Channels. 2016;**10**:101-110

[20] Brette F, Luxan G, Cros C, Dixey H, Wilson C, Shiels HA. Characterization of isolated ventricular myocytes from adult zebrafish (*Danio rerio*). Biochemical and Biophysical Research Communications. 2008 Sep 12;**374**(1):143-146

[21] Vornanen M. Electrical excitability of the fish heart and its autonomic regulation. In: Gamperl KA, Gillis TE, Farrell PA, Brauner CJ, editors. Fish Physiology. The Cardiovascular Physiology. Morphology, Control and Function. London, UK: Elsevier; 2017. pp. 99-153

[22] Vornanen M. The temperature-dependence of electrical excitability of fish heart. The Journal of Experimental Biology. 2016;**219**:1941-1952

[23] Barrionuevo WR, Burggren WW. O_2 consumption and heart rate in developing zebrafish (*Danio rerio*): Influence of temperature and ambient O_2. The American Journal of Physiology. 1999;**276**:R505-R513

[24] Haverinen J, Vornanen M. Temperature acclimation modifies sinoatrial pacemaker mechanism of the rainbow trout heart. The American Journal of Physiology. 2007 Feb; **292**:R1023-R1032

[25] Haverinen J, Vornanen M. Temperature acclimation modifies Na^+ current in fish cardiac myocytes. The Journal of Experimental Biology. 2004;**207**:2823-2833

[26] Vornanen M, Ryökkynen A, Nurmi A. Temperature-dependent expression of sarcolemmal K^+ currents in rainbow trout atrial and ventricular myocytes. The American Journal of Physiology. 2002;**282**:R1191-R1199

[27] Grandi E, Pasqualini FS, Bers DM. A novel computational model of the human ventricular action potential and Ca transient. Journal of Molecular and Cellular Cardiology. 2010;**48**(1):112-121

[28] Chopra SS, Stroud DM, Watanabe H, Bennett JS, Burns CG, Wells KS, et al. Voltage-gated sodium channels are required for heart development in zebrafish. Circulation Research. 2010 Apr 30;**106**(8):1342-1350

[29] Haverinen J, Hassinen M, Vornanen M. Fish cardiac sodium channels are tetrodotoxin sensitive. Acta Physiologica. 2007;**191**(3):197-204

[30] Vornanen M, Hassinen M, Haverinen J. Tetrodotoxin sensitivity of the vertebrate cardiac Na^+ current. Marine Drugs. 2011;**9**:2409-2422

[31] Catterall WA, Perez-Reyes E, Snutch TP, Striessnig J. International Union of Pharmacology. XLVIII. Nomenclature and structure-function relationships of voltage-gated calcium channels. Pharmacological Reviews. 2005 December 01;**57**(4):411-425

[32] McDonald TF, Pelzer S, Trautwein W, Pelzer DJ. Regulation and modulation of calcium channels in cardiac, skeletal, and smooth muscle cells. Physiological Reviews. 1994; **74**(2):365-507

[33] Beuckelmann DJ. Contributions of Ca^{2+}-influx via the L-type Ca^{2+}-current and Ca^{2+}-release from the sarcoplasmic reticulum to $[Ca^{2+}]_i$-transients in human myocytes. Basic Research in Cardiology. 1997;**92**(Suppl. 1):105-110

[34] Mewes T, Ravens U. L-type calcium currents of human myocytes from ventricle of non-failing and failing hearts and from atrium. Journal of Molecular and Cellular Cardiology. 1994;**26**(10):1307-1320

[35] Gaborit N, Le Bouter S, Szuts V, Varro A, Escande D, Nattel S, et al. Regional and tissue specific transcript signatures of ion channel genes in the non-diseased human heart. The Journal of Physiology. 2007 Jul 15;**582**(Pt 2):675-693

[36] Haverinen J, Dash SN, Hassinen M, Vornanen M. The large T-type Ca2+ current of the zebrafish ventricular myocytes is mainly produced by α1G (Ca_v3.1) channels. 2017. (submitted)

[37] Baker K, Warren KS, Yellen G, Fishman MC. Defective "pacemaker" current (I_h) in a zebrafish mutant with a slow heart rate. Proceedings of the National Academy of Sciences of the United States of America. 1997;**94**:4554-4559

[38] Hibino H, Inanobe A, Furutani K, Murakami S, Findlay I, Kurachi Y. Inwardly rectifying potassium channels: Their structure, function and physiological role. Physiological Reviews. 2010;**90**:291-366

[39] Hassinen M, Haverinen J, Hardy ME, Shiels HA, Vornanen M. Inward rectifier potassium current (I_{K1}) and Kir2 composition of the zebrafish (*Danio rerio*) heart. Pflügers Archiv. 2015;**467**:2437-2446

[40] ten Tusscher KHWJ, Noble D, Noble PJ, Panfilov AV. A model for human ventricular tissue. The American Journal of Physiology. 2004;**286**(4):H1573-H1589

[41] Priebe L, Beuckelmann DJ. Simulation study of cellular electric properties in heart failure. Circulation Research. 1998 Jun 15;**82**(11):1206-1223

[42] Ravens U, Wettwer E. Ultra-rapid delayed rectifier channels: molecular basis and therapeutic implications. Cardiovascular Research. 2011 March 01;**89**(4):776-785

[43] Scholz EP, Niemer N, Hassel D, Zitron E, Burgers HF, Bloehs R, et al. Biophysical properties of zebrafish ether-a-go-go related gene potassium channels. Biochemical and Biophysical Research Communications. 2009 Apr 3;**381**(2):159-164

[44] Leong IUS, Skinner JR, Shelling AN, Love DR. Identification and expression analysis of kcnh2 genes in the zebrafish. Biochemical and Biophysical Research Communications. 2010 Jun 11;**396**(4):817-824

[45] Peal DS, Mills RW, Lynch SN, Mosley JM, Lim E, Ellinor PT, et al. Novel chemical suppressors of long QT syndrome identified by an in vivo functional screen. Circulation. 2011 January 04;**123**(1):23-30

[46] Gerdes AM, Capasso JM. Structural remodeling and mechanical dysfunction of cardiac myocytes in heart failure. Journal of Molecular and Cellular Cardiology. 1995; **27**(3):849-856

[47] Jopling C, Sleep E, Raya M, Martí M, Raya A, Belmonte JCI. Zebrafish heart regeneration occurs by cardiomyocyte dedifferentiation and proliferation. Nature. 2010; **464**(7288):606-609

[48] Bovo E, Dvornikov AV, Mazurek SR, de Tombe PP, Zima AV. Mechanisms of Ca^{2+} handling in zebrafish ventricular myocytes. Pflügers Archiv. 2013 Dec;**465**(12):1775-1784

[49] Fabiato A. Calcium-induced release of calcium from the cardiac sarcoplasmic reticulum. The American Journal of Physiology. 1983;**245**:C1-C14

[50] Piacentino III V, Weber CR, Chen X, Weisser-Thomas J, Margulies KB, Bers DM, et al. Cellular basis of abnormal calcium transients of failing human ventricular myocytes. Circulation Research. 2003;**92**(6):651-658

[51] Zhang PC, Llach A, Sheng XY, Hove-Madsen L, Tibbits GF. Calcium handling in zebrafish ventricular myocytes. The American Journal of Physiology. 2011 Jan;**300**(1):R56-R66

[52] Beuckelmann DJ, Näbauer M, Erdmann E. Intracellular calcium handling in isolated ventricular myocytes from patients with terminal heart failure. Circulation. 1992 Mar; **85**(3):1046-1055

[53] Vornanen M. Temperature and Ca^{2+} dependence of $[^{3}H]$ryanodine binding in the burbot (*Lota lota* L.) heart. The American Journal of Physiology. 2006;**290**(2):R345-R351

[54] O'Hara T, Rudy Y. Quantitative comparison of cardiac ventricular myocyte electrophysiology and response to drugs in human and nonhuman species. The American Journal of Physiology. 2012 Mar 1;**302**(5):H1023-H1030

[55] Jost N, Virág L, Comtois P, Ördög B, Szuts V, Seprényi G, et al. Ionic mechanisms limiting cardiac repolarization reserve in humans compared to dogs. Journal of Physiology. 2013 September 01;**591**(17):4189-4206

[56] Wright SN, Wang SY, Kallen RG, Wang GK. Differences in steady-state inactivation between Na channel isoforms affect local anesthetic binding affinity. Biophysical Journal. 1997 Aug;**73**(2):779-788

[57] Perry M, Sanguinetti MC. A single amino acid difference between ether-a-go-go-related gene channel subtypes determines differential sensitivity to a small molecule activator. Molecular Pharmacology. 2008 Apr;**73**(4):1044-1051

[58] Hassinen M, Haverinen J, Vornanen M. Molecular basis and drug sensitivity of the delayed rectifier (I_{Kr}) in the fish heart. Comparative Biochemistry and Physiology. C. 2015 Oct-Nov;**176-177**:44-51

[59] Kramer J, Obejero-Paz CA, Myatt G, Kuryshev YA, Bruening-Wright A, Verducci JS, et al. MICE models: Superior to the HERG model in predicting Torsade de Pointes. Scientific Reports. 2013;**3**:2100

[60] Redfern WS, Carlsson L, Davis AS, Lynch WG, MacKenzie I, Palethorpe S, et al. Relationships between preclinical cardiac electrophysiology, clinical QT interval prolongation and torsade de pointes for a broad range of drugs: evidence for a provisional safety margin in drug development. Cardiovascular Research. 2003 Apr;**58**(1, 1):32-45

[61] Colatsky T, Fermini B, Gintant G, Pierson JB, Sager P, Sekino Y, et al. The comprehensive in vitro proarrhythmia assay (CiPA) initiative—Update on progress. Journal of Pharmacological and Toxicological Methods. 2016;**81**:15-20

[62] Choi BR, Burton F, Salama G. Cytosolic Ca^{2+} triggers early afterdepolarizations and Torsade de Pointes in rabbit hearts with type 2 long QT syndrome. The Journal of Physiology. 2002;**543**(2):615-631

[63] Kihara Y, Morgan JP. Intracellular calcium and ventricular fibrillation. Studies in the aequorin-loaded isovolumic ferret heart. Circulation Research. 1991;**68**(5):1378-89

[64] Poss KD, Wilson LG, Keating MT. Heart regeneration in zebrafish. Science. 2002;**298**: 2188-2190

[65] Martinez ML, Heredia MP, Delgado C. Expression of T-type Ca^{2+} channels in ventricular cells from hypertophied rat hearts. Journal of Molecular and Cellular Cardiology. 1999;**31**:1617-1625

[66] Izumi T, Kihara Y, Sarai N, Yoneda T, Iwanaga Y, Inagaki K, et al. Reinduction of T-type calcium channels by endothelin-1 in failing hearts in vivo and in adult rat ventricular myocytes in vitro. Circulation. 2003 Nov 18;**108**(20):2530-2535

[67] Lyon AR, MacLeod KT, Zhang Y, Garcia E, Kanda GK, Lab MJ, et al. Loss of T-tubules and other changes to surface topography in ventricular myocytes from failing human and rat heart. Proceedings of the National Academy of Sciences of the United States of America. 2009 Apr 21;**106**(16):6854-6859

[68] Qu Y, Boutjdir M. Gene expression of SERCA2a and L-and T-type Ca channels during human heart development. Pediatric Research. 2001;**50**(5):569-574

[69] Kim HD, Kim DJ, Lee IJ, Rah BJ, Sawa Y, Schaper J. Human fetal heart development after mid-term: Morphometry and ultrastructural study. Journal of Molecular and Cellular Cardiology. 1992;**24**(9):949-965

[70] Hu N, Yost HJ, Clark EB. Cardiac morphology and blood pressure in the adult zebrafish. Anatomical Record. 2001;**264**:1-12

[71] Guyton AC. Textbook of Medical Physiology. Narwalk, USA: Sounders Company; 1981. pp. 1-1074

[72] Li GR, Feng J, Yue L, Carrier M, Nattel S. Evidence for two components of delayed rectifier K^+ current in human ventricular myocytes. Circulation Research. 1996 Apr;**78**(4):689-696

[73] Wettwer E, Amos GJ, Posival H, Ravens U. Transient outward current in human ventricular myocytes of subepicardial and subendocardial origin. Circulation Research. 1994 Sep;**75**(3):473-482

[74] Nelson DA, Benson ES. On the structural continuities of the transverse tubular system of rabbit and human myocardial cells. The Journal of Cell Biology. 1963 Feb;**16**:297-313

[75] Bassani JWM, Bassani RA, Bers DM. Relaxation in rabbit and rat cardiac cells: Species-dependent differences in cellular mechanisms. The Journal of Physiology. 1994; **476**:279-293

[76] Bers DM. Cardiac excitation-contraction coupling. Nature. 2002;**415**:198-205

[77] Franz MR, Swerdlow CD, Liem LB, Schaefer J. Cycle length dependence of human action potential duration in vivo. Effects of single extrastimuli, sudden sustained rate acceleration and deceleration, and different steady-state frequencies. The Journal of Clinical Investigation. 1988 Sep;**82**(3):972-979

2

The Monitoring and Assessment of Cd²⁺ Stress Using Zebrafish (*Danio rerio*)

Zongming Ren and Yuedan Liu

Abstract

Pollution on the Earth is ubiquitous across ecosystems from the land to the ocean. Various sources contribute to pollution including industrial (e.g. chemicals), agricultural (e.g. pesticides) and domestic (e.g. transportation) pollutants' ecosystems and substrate environment (e.g. contamination in water). The extensive use of chemicals in agriculture, forests and wetlands may impair biological communities. Due to the lack of target specificity, these chemicals can cause severe and persistent toxic effects on nontarget aquatic species, including bacteria, invertebrates and vertebrates. Different degrees of biological response have been presented according to intensities of different chemicals. The cadmium (Cd^{2+}) contamination in aquatic environment has attracted more and more attention due to its toxic characteristics, for example, accumulation in environment, nondegradability and the potential threat to the ecosystem. Knowledge and understanding of these conditions have led to the development of new monitoring and assessment technologies based on biological and chemical methods. This chapter covers new monitoring technologies and environment assessment of Cd^{2+} stress using zebrafish (*Danio rerio*), which include the behaviour responses, metabolism and electrocardiogram (ECG).

Keywords: zebrafish, behaviour responses, metabolism, electrocardiogram (ECG), cadmium stress

1. Introduction

With the development of industrialization, the efforts of decreasing pollutants from the natural environment cannot satisfy the fundamental requirement of health and safety due to the large and increasing amounts of waste materials [1]. Environmental pollution has become an increasingly serious problem in recent years [2, 3], and water pollution has caused more and more concern. The extensive use of chemicals in agriculture, forests and wetlands may impair biological communities [4, 5].

Due to the lack of target specificity, most contaminants can cause severe and persistent toxic effects on non-target aquatic species, including invertebrates and vertebrates [6, 7]. Knowledge and understanding of these conditions have led to the development of new monitoring, analysis and assessment technologies based on biological, physical and chemical methods [8–10]. Campanella et al. [11] achieved the monitoring of the evolution of photosynthetic oxygen (O_2) and the detection of alterations due to toxic effects caused by environmental pollutants using a sensor by coupling a suited algal bioreceptor to an amperometric gas diffusion electrode. Glasgow et al. [12] illustrated that real-time remote monitoring and sensing technology which is a more important tool for evaluation of water quality. Zhang et al. [13] developed an online behaviour monitoring system to assess the toxic effects of carbamate pesticides on medaka (*Oryzias latipes*), a small fish native to East Asia. Alexakis [14] assessed the water samples based on chemical indexes such as sodium adsorption ratio, sodium percentage and residual sodium carbonate. These methods greatly contribute to environmental stress assessment and water resource management.

Heavy metal pollution caused worldwide attention due to their bioaccumulation. Cadmium (Cd) is one of the most easily accumulated toxic substances in humans and organisms [15]. Cd pollution commonly occurs in food and environment (e.g. water, air), and also Cd has a long half-life period in human and animal bodies. Hansen has investigated that rats accumulated Cd^{2+} at a greater rate firstly and the kidney toxicity due to Cd^{2+} needed 6–8 weeks [16]. The accumulation of cadmium could be observed in different organs in fish [17] (**Figure 1**).

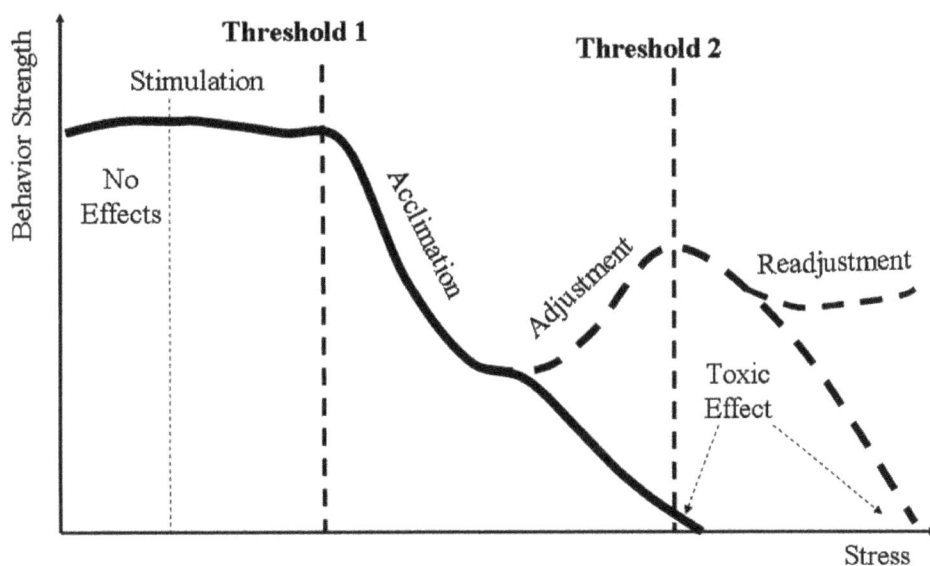

Figure 1. Step-wise behaviour responses according to previous reports: The model showed a series of different behaviours including "No effect," "Acclimation," "Adjustment (Readjustment)" and "Toxic effect". Threshold, the point at which a physiological effect begins to be specifically manifested on the individuals after exposure to stressors, decided the behaviour responses.

In the ecological assessment of environmental stress, many methods in different levels (molecular, cell, physiological and individual) are used [18]. Behaviour responses to toxic effects have been specifically observed [19, 20]. It has been reported that *Daphnia magna*, medaka (*Oryzias latipes*) and rare minnow (*Gobiocypris rarus*) are aquatic homeostatic organisms [21, 22] showing evident step-wise behaviour response, including "No effect", "Acclimation", "Adjustment (Readjustment)" and "Toxic effect" [23–25]. Wang et al. [26] reported a sequence of intoxication and recovery processes through data transformation (i.e. integration) at the time progressed after exposure to toxic chemicals and suggested that the behaviour response is a good indicator for aquatic organisms to assess the water quality.

As an endpoint in the physiological level, organisms' metabolism plays an important role in the environmental stress assessment. Metabolism is the fundamental process of organisms, and it is related to energy assimilation, transformation and allocation that strongly influence the rate of individual growth and reproduction [27]. Standard metabolic rate (SMR), which is the minimum metabolic level of fish in the state of rest and starvation, means the metabolism level of organisms, which could be analysed directly by oxygen consumption (OC) monitoring system [28]. Respiration is an important physiological characteristic of the metabolic activities of fish, which can reflect the adaptation of fish to the external environment. Metabolism in the characteristic of SMR is ordinarily equal to OC [29] on the condition of no food supplied during the monitoring of fish respiration. Usually, there are many factors that can affect OC of fish including water temperature, body mass, body size, dioxide oxygen, sanity, atmospheric pressure, injury and disease [28].

Heart ECG provides a chart that represents the electrical activity of the heart, and it also provides a time voltage of the heartbeat [30]. The notation of ECG waveform suggested by Einthoven [31] is still in use today and has been used to detect and monitor disease in different animals [32–34]. ECG waves (P, Q, R, S and T) and the time intervals (PR, QRS, ST and QT) have been used to differentiate healthy and diseased fish [35]. Therefore, fish ECG can be a good tool to assess water quality.

Zebrafish (*Danio rerio*), whose ether-a-go-go-related gene (zERG) which has high similarities in the protein sequence with the human gene (hERG) [35] has been frequently used as a representative in the toxicologic assessment of chemicals [36], providing sensitive, economical, practical and biological monitors for aquatic pollutants [37, 38]. First, it is economic to use zebrafish to assess water quality. Second, zebrafish can be a tool as good as higher vertebrates on toxic testing. Third, the use of zebrafish with other fish can realize the potential toxicity analysis [39]. Therefore, new monitoring technologies based on the behaviour responses, metabolism and electrocardiogram (ECG) of zebrafish can be applied to realize the monitoring and assessment of Cd^{2+} stress.

2. Methods

In the recording of the behaviour responses, metabolism and ECG, room temperature is controlled at $26 \pm 2°C$ with a photoperiod of 16 h light and 8 h dark. Nonchlorinated water (hardness based on $CaCO_3$: 250 ± 25 mg/L, pH 7.8 and temperature 26 ± 2°C) is used. No food is provided to test organisms during the assessment.

2.1. The recording of behaviour responses

An online monitoring system, built in the Research Center for Eco-Environmental Science, Chinese Academy of Sciences [40], is used to analyse the continuous swimming behaviour (**Figure 2**). Behaviour data are collected by behaviour sensors, which are made up of two pairs of electrodes that sent a high-frequency signal of alternating current by one pair and then received another. The behaviour strength is sampled automatically every second, and the average behaviour strength data are taken twice an hour in the first 2 h. Behaviour strength that changes from 0 (losing the ability of movement) to 1 (full behaviour expression) is applied to represent the differences of behavioural responses [13].

Test zebrafish were placed in behaviour sensors (10 cm long, 7 cm in diameter). These sensors were closed off on both sides with 250 μm nylon nets, and three replicates per concentration were used. Flow rate of each test channel was controlled about 2 L/h. Temperature and light conditions were the same as stock rearing. The flow rate in each test chamber was controlled to about 1 L/h in the online mixing system, and then the flow rate in the online monitoring system of behaviour strength was about 2 L/h.

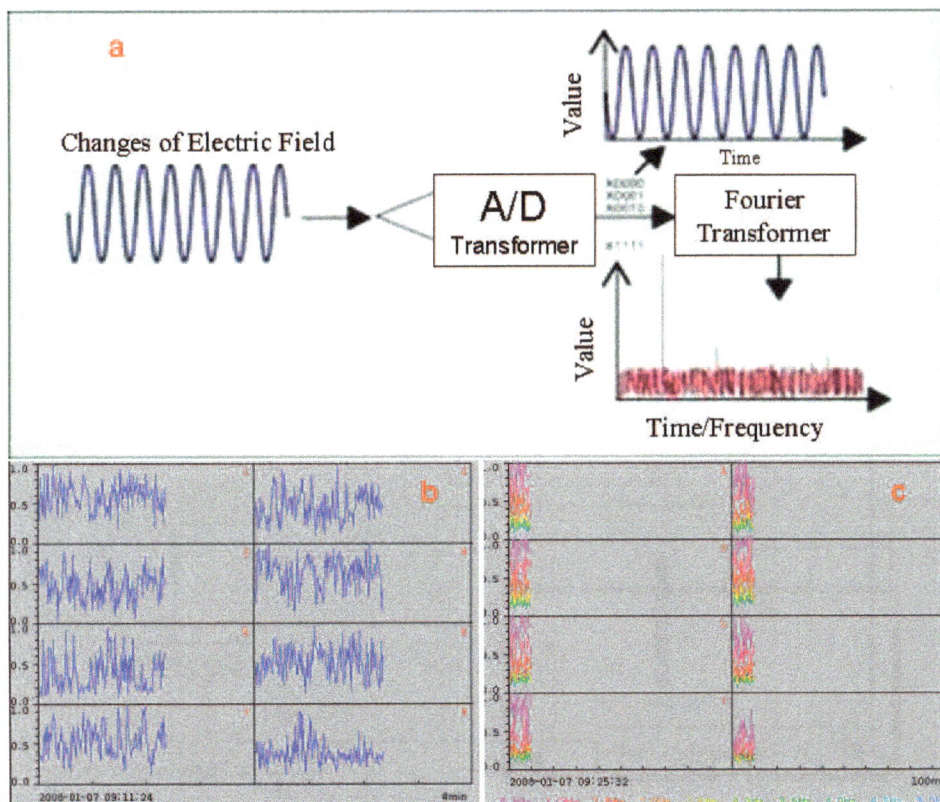

Figure 2. Signal acquisition and transmission of the behaviour responses: (a) the signal acquisition, (b) the normal signal analysis, and (c) the signal analysis after Fast Fourier Transform.

2.2. The recording of metabolism

An online monitoring system based on OC is used to record the metabolism of zebrafish (**Figure 3**). The metabolism monitoring system is made up of water tank, flow-through sensor, peristaltic pump, three-way valve, digital control unit, data acquisition unit, dissolved oxygen (DO) sensor, thermometer, pressure meter and water tubes.

Before the assessments, test zebrafish volumes are measured by 5 ml graduated glass cylinder (Va), and the body mass of test zebrafish is weighted as follows: A beaker (50 ml) with 10 ml water is weighted by a precision electronic balance (FA2204N, Shanghai Jinghai Instrument), which is regarded as a baseline weight. Then, test individuals are put into the beaker to get the wet weight (m). Test zebrafish are placed into the flow-through sensors (200 ml, Vr) with nylon nets (250 µm) at both sides to prevent test zebrafish running into water tubes. The water flow rate of the sensor is controlled by a peristaltic pump at approximately 2 L/h. The whole cycle of the three-way valve is designed as 150 s flushing phase (to ensure DO in test chamber is enough for test individuals) after 300 s circulation phase (to make measurement, regarded as t). During the circulation mode, the three-way valve is energized to close the loop. In this mode, the water flows from the flow-through sensor to the DO sensor. When the organisms breathe, the DO sensor can realize the measurement. In this phase, water tubes are installed in the correct direction, and all connections are sealed completely to ensure no external oxygen enters the system. During the flushing mode (dotted lines), the three-way valve will pump water (black dot) from water tank (brown box) to test chamber, and the water body will flow through chamber into water tank again (red dot). The flushing and circulation cycle are repeated until the end of the experiment. Oxygen concentrations of both inflow and outflow water are detected by DO sensor, and then these data are recorded as DO_i and DO_o. Absolute room pressure is provided to ensure the stability of DO in the system. Prior to the start of the experiment, all sensors are warmed up for 5 min, and then the OC data begin to be collected. Water temperature is recorded to remain constant through the experiments.

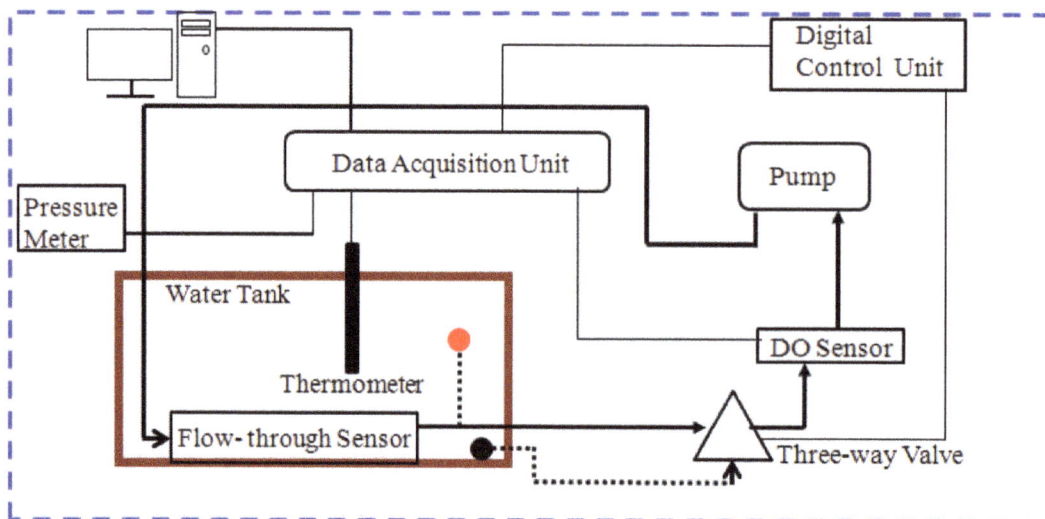

Figure 3. Online monitoring systems of metabolism based on oxygen consumption (OC).

DO (mg/L) data in the metabolism monitoring system are converted to OC using Eq. (1):

$$VO_2 = \frac{(DO_i - DO_o) \times (Vr - Va)}{m \times t}$$

(1)

in which VO_2: OC of test zebrafish (mg/kg/h); DO_i: DO concentration of water body flowing into SMR sensors (mg/L); DO_o: DO concentration of water body flowing out SMR sensors (mg/L); V_r: the volume of flow-through sensor (L) and it is 0.2 L for zebrafish; V_a: test zebrafish volume (L); m: test zebrafish mass (kg); and t: time of circulation cycle (h).

2.3. The recording of ECG

The ECGs are detected by RM-6240C Multichannel Physiological Signal Acquisition and Processing System (ChengDu Instrument Factory, China). During ECG acquisition, zebrafish

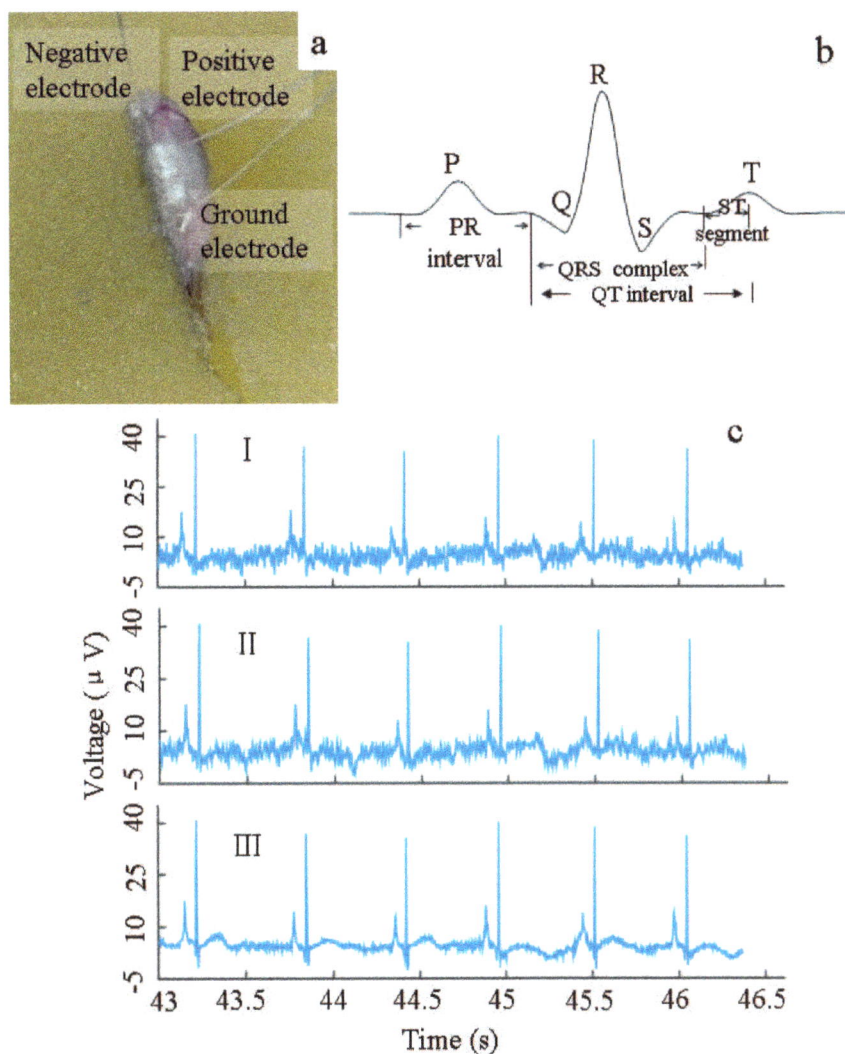

Figure 4. The acquisition and the analysis of zebrafish ECG.

are fixed in a damp sponge with the ventral side exposed for electrode placement (**Figure 4a**). Two micro-electrodes are positioned at 90° to the animal's epidermis near the heart and inside the medial pectoral fin; the positive electrode is on the right side and the negative electrode is at a relatively lower position on the left side. The grounding electrode is closer to the top of the pelvic fins as presented in **Figure 4a**. The relative distance between grounding electrode and other two electrodes is 8–10 mm. All electrodes are inserted into the skin to approximately 1 mm in depth. The exact position of the positive and negative electrodes was adjusted slightly to obtain the maximum voltage signals. ECG signals are characterized by the following parameters: waves P, Q, R, S, T and intervals PR, QRS, ST, QT (**Figure 4b**). ECG signals of each zebrafish are obtained over a 3 min period. Baseline ECG signals are recorded 24 h before assessment. The positive and negative electrodes are attached to the atrioventricular junction of the zebrafish, and the ECG signals are recorded at a sampling frequency of 4 kHz with the time constant 0.02 s [**Figure 4c(I)**]. The recorded signals were digitally processed by MATLAB (MATLAB R2009a, The MathWorks Inc., Natick, MA, USA). First, a zero-phase digital filter is used [**Figure 4c(II)**], and then the wavelet decomposition is acquired by performing a 10-level 1D wavelet analysis using coif5 wavelet. With the wavelet decomposition structure and the coif5 wavelet, the coefficients of the filtered signals are further reconstructed. The de-noised signals [**Figure 4c(III)**] are finally obtained from the above-reconstructed wavelet decomposition structure using the principle of Stein's unbiased risk; soft threshold, level-dependent estimation of level noise and coif5 wavelet at level 10 are used to analyse the ECG changes.

3. Results and discussion

3.1. The behaviour responses

According to the acute toxicity experiment following the guidelines for fish acute toxicity from OECD 203 [41], 48 h median lethal concentration (LC_{50}–48 h) of cadmium chloride ($CdCl_2$) to zebrafish after probit analysis in MATLAB with 95% confidence interval is 42.6 mg/L. The LC_{50}–48 h value (42.6 mg/L) is regarded as 1.0 TU, and then the exposure concentrations are 4.26, 42.6 and 85.2 mg/L in 0.1, 1.0 and 2.0 TU $CdCl_2$ treatments, respectively.

The behaviour responses based on behaviour strength (BS) of zebrafish are shown in **Figure 5**. The average BS in control kept about 0.8, and the higher the chemical concentrations, the lower is the behaviour strength. These results suggest that the toxic effects of $CdCl_2$ on the BS of zebrafish are closely related to the concentrations, which was defined as dose-effect relationship [42]. It is noteworthy that lower BS values could be observed in the dark period at 13–21 and 37 and 45 h in all treatments. Considering the ups and downs in BS curves matched to photo and scoot phase, this may suggest a possibility of circadian rhythm in test organisms after being exposed to the chemical [43].

The behaviour activity (BA) as integrated BS was obtained by linear regression (**Figure 6**). If BA is in the positive range, the BS would be higher than the accumulated average of BS and could be considered as in active state [44]. The BA values were mostly negative at initial stage. The crossing times of BA between positive and negative values were commonly observed which show the circadian rhythms. The control group showed the high value between 21 and 37 h,

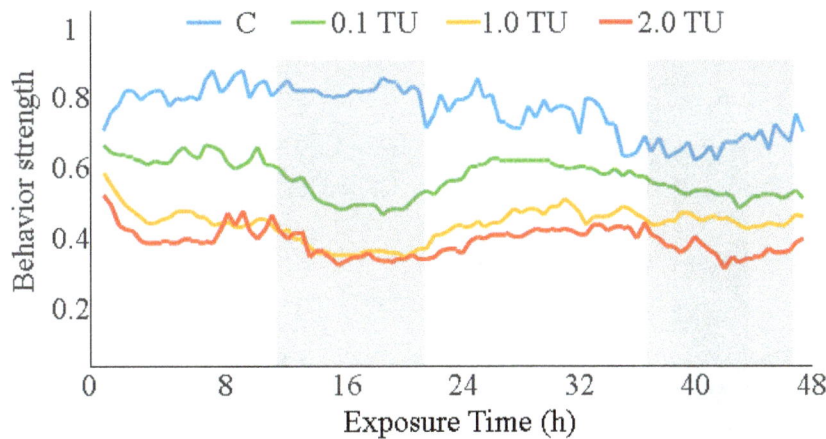

Figure 5. Behaviour responses of zebrafish in the treatments of $CdCl_2$. Grey shadow means the dark periods as the experiment started from 9 o'clock in the morning.

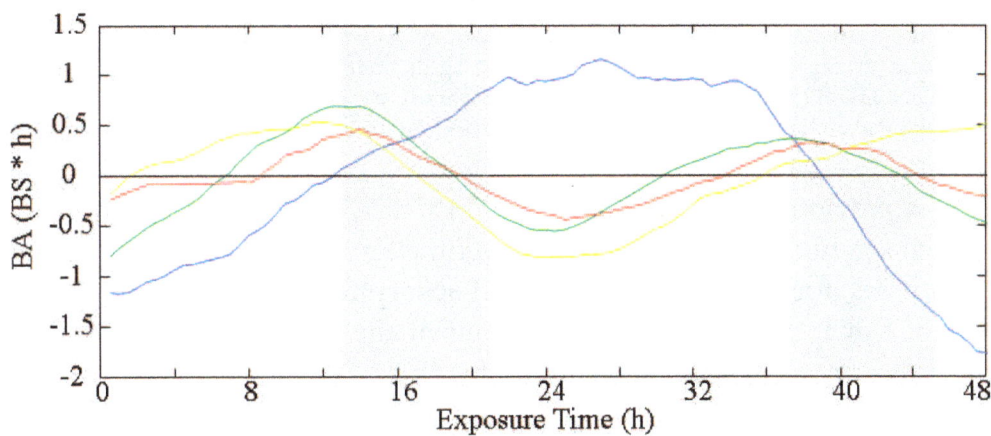

Figure 6. BA values for $CdCl_2$ across different treatments. Control (blue), 0.1TU (green), 1.0 TU (yellow) and 2.0 TU (red). Shadows mean dark periods.

which matched photophase. However, BS values were not high at either previous (0–13 h) or next (45–48 h) photoperiod. Remarkably, the circadian rhythms appeared in different treatments, being substantially different with the control group. Except the last phase with 1.0 TU, rhythmic activity was clearly observed at dark phases, 13–21 and 37–45 h, which indicated that the pollutant is a stimulating agent to resume the circadian system in test organisms. BS was higher at scoot phase. It was noteworthy that the rhythm was disrupted with 1.0 TU at the end of the experiment, whereas the lower (0.1 TU) and higher (2 TU) levels showed clear rhythms. The circadian rhythms of the test group were quite different from the control group, indicating that the environmental stress stimulated the biological clock of zebrafish and made it more clear in pollutants than in control [45], which is consistent with the result of **Figure 5**.

3.2. The changes of metabolism

As presented in **Figure 7**, the exposure of $CdCl_2$ significantly affected the overall oxygen consumption (OC) of zebrafish ($p < 0.05$). OC at 0.1 TU (463.11 ± 44.21 mg/kg/h) was significantly ($p < 0.05$) lower than the control (617.39 ± 30.48 mg/kg/h) and higher than 1.0 TU (314.40 ± 40.04 mg/kg/h) and 2.0 TU (229.07 ± 28.66 mg/kg/h). These results could be

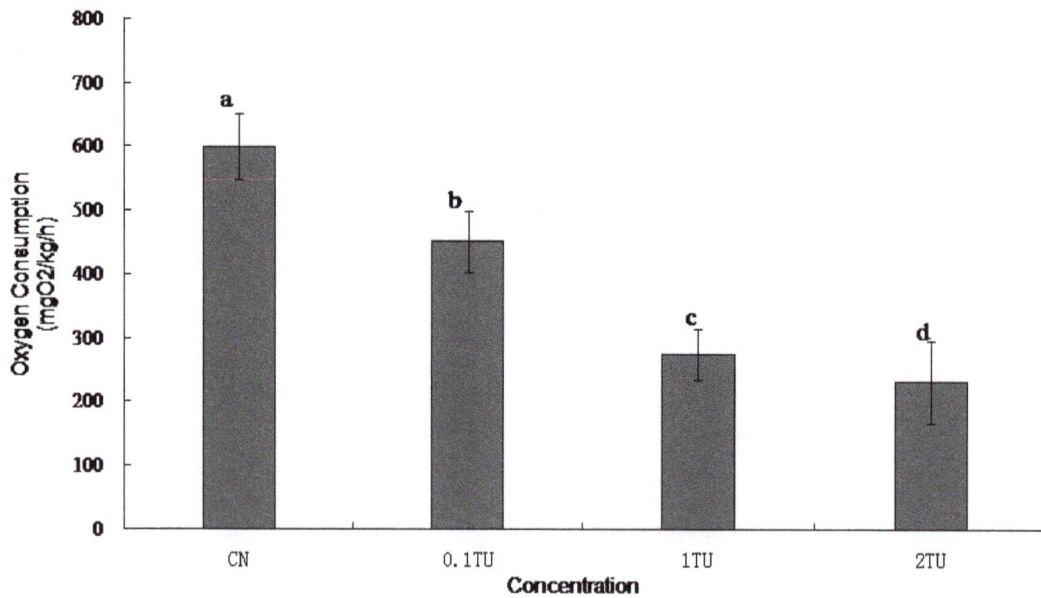

Figure 7. Changes in average oxygen consumption (mgO2/kg/h) after exposure to $CdCl_2$ across different toxic concentrations, 0TU, 0.1TU, 1TU and 2TU. The bars in the histogram represent the mean ± SD of oxygen consumption in different TUs. Different alphabets on the bar represent statistical significance based on multiple comparison ($p < 0.05$).

supported by a previous report on the effects of $CdCl_2$ on the respiration of fresh water crab [46], which is a fundamental physiological function in organisms that can affect other metabolic processes, for example, feeding, food absorption and excretion [47]. In the toxic environment of $CdCl_2$, the internal environment of the cell changed, and all the related energy processes were changed correspondingly [48].

Figure 8 shows the continuous changes of zebrafish OC during 48 h exposure of $CdCl_2$, which suggested that the total tendency of the continuous results had the same order as presented in **Figure 7**. In control, a high respiration rate was maintained at about 617.39 mg/kg/h during the observation period. In different treatments, the observed levels of OC decreased with the increase in concentrations, for example, OC was approximately 500 mg/kg/h in 0.1 TU, 300 mg/kg/h in 1.0 TU and 200 mg/kg/h in 2.0 TU.

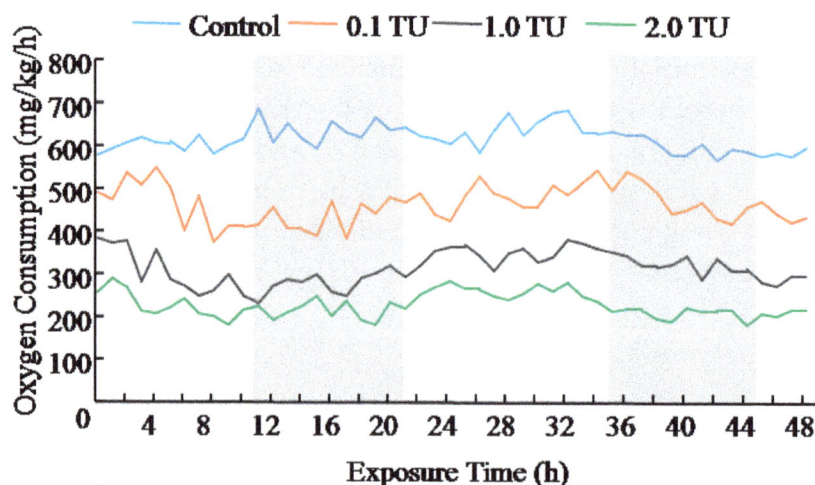

Figure 8. Continuous changes of zebrafish OC during 48 h exposure in differentCdCl₂treatments.

During 48 h exposure, OC in most treatments had some circadian rhythms according to the average values in different periods (**Table 1**) because OCs were significantly ($p < 0.05$) higher during photo phase (D1 and D2) than scoot period (N1 and N2).

When zebrafish were exposed to 0.1 TU and 1.0 TU $CdCl_2$, OCs were significantly ($p < 0.05$) different between sooct phase and photo phase. This indicated that the response of zebrafish to pollutants was stronger and had great effects on diurnal variation during 48 h exposure. There was a greater fluctuation of OC in $CdCl_2$ exposures than in the control. In most treatments, higher values were observed during photo phases (0.1, 1 and 2 TU). At the beginning, OC decreased until the 11th h and then recovered during 21–36 h. In the end, OC decreased at all concentrations during the exposure of $CdCl_2$ in 36–48 h.

3.3. The changes of ECG

Figure 9 shows the average values of zebrafish ECG characteristics in different treatments. The values of all exposure time points during the 48-h exposure are assigned to the corresponding concentration groups. After exposure to $CdCl_2$, the amplitudes of all waves (P, Q, R, S and T) and interval durations (PR, QRS, ST and QT) showed some difference with significance ($p < 0.05$ or $p < 0.01$). In the control group, the amplitudes of all waves were the largest, and they were almost the smallest in different intervals. Overall, the amplitudes of waves tended to show negative relationships, and intervals showed positive relationship with $CdCl_2$ concentrations. The changes of waves that Q, T, QRS and QT intervals showed clearly observed dose-effect relationship. There were some exceptions: S and ST did not show an obvious dose-effect relationship, and R showed a reverse effect to other waves. These results suggested that it was not sufficient to analyse the toxic effects of environmental stress on zebrafish ECG alone, depending on the average values of these characteristics.

To assess water quality using zebrafish ECG characteristics, the continuous changes based on the de-noised ECG signals were applied (**Figure 10**). The difference of zebrafish ECG characteristics after two-way ANOVA is shown in **Table 2**. Some characteristics (Q, R, S and T) showed some significant difference ($p < 0.05$ or $p < 0.01$) at different exposure times in the control group (**Figure 10b–e**), but the differences of P, PR, QRS, ST and QT showed no significant changes, which suggests that P, PR, QRS, ST and QT can serve as normal control in the analysis of $CdCl_2$ toxic effects on zebrafish ECG (**Figure 10a**, f, g, h, i).

Time periods	Control	0.1 TU	1.0 TU	2.0 TU
D1	610.47 ± 25.69*aBb	460.38 ± 52.88	310.53 ± 52.92	236.69 ± 32.28*b
N1	632.18 ± 23.31*Ab	436.54 ± 36.44*B	285.47 ± 21.01*Bb	214.81 ± 22.22*B
D2	633.20 ± 33.03*Ab	488.10 ± 37.48*a	342.35 ± 27.01*ab	246.67 ± 23.02*ab
N2	590.88 ± 17.78*AaB	458.77 ± 30.42	309.99 ± 22.47*aB	207.22 ± 12.83*AB

D1, the first 0 day results, from 0 to 11 h and from 21 to 24 h; D2, the second-day results, from 24 to 36 h and from 46 to 48 h; N1, the first-night results, from 11 to 21 h; N2, the second-night results, from 36 to 46 h exposure. Data are shown as M ± S.D.* $p < 0.05$, A, B, a and b mean the significant difference with D1, D2, N1 and N2, respectively.

Table 1. OC of zebrafish in different treatments during both photo phase (D) and scoot period (N).

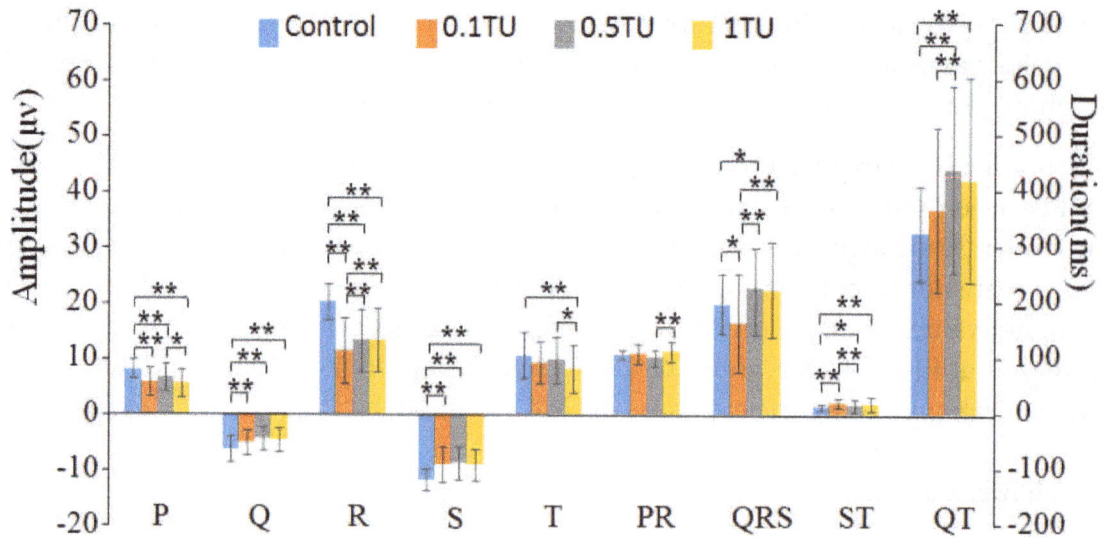

Figure 9. Comparison of zebrafish ECG characteristics after treated with $CdCl_2$. The unit of wave amplitude (P, Q, R, S and T) is μv, the unit of duration (PR interval, ST segment, QRS complex and QT interval) is ms. Vertical bars indicate mean values and the solid lines are standard deviation. The data of every ECG parameter are representative of at least four independent experiments. — represents two experimental groups indicating the ends of a line segment. $*p < 0.05$, $**p < 0.01$.

The observed continuous changes of ECG suggested that signals of P, Q, T and PR were disordered, in which the data of amplitudes and durations showed no clear regularity depending on either exposure time or concentrations. According to the analysis of the tendency (**Table 2** and **Figure 10**), R, QRS, ST and QT in different treatments showed dose dependency; however, due to some fluctuation changes at 8, 16 and 32 h, ST and QT did not show clear time dependence.

Overall, QRS, ST and QT showed prolonged effects after $CdCl_2$ exposure and wave R showed a decrease in amplitude. A prolonged ST reflects delayed repolarization of ventricular function. As it is reported, ST depression hysteresis could offer a substantially better diagnostic accuracy for coronary artery disease [49], and ST elevation could be induced by $CdCl_2$ [50]. A wide QRS reflecting left-sided intra-ventricular conduction delay and prolonged QRS reflects delayed depolarization of ventricular function too. Prolonged QT reflects delayed repolarization of action potential [51]. $CdCl_2$ is highly toxic to the cardiomyocytes, characterized by lengthening the repolarization phase of the cardiac action potential, manifested as prolongation of the QT on the surface ECG and predisposition to special arrhythmia [35]. Therefore, the QT prolonging suggested that $CdCl_2$ induced bradycardia and arrhythmia in zebrafish. The R wave voltage changes were observed with doxorubicin (DXR) treatment. On the cellular level, DXR treatment led to a decrease in V_{max}, with little increase or no change in resting potential and a marked prolongation in action potential duration at 50–75% repolarization levels [52].

To specify the role of ECG characteristics in the assessment of $CdCl_2$ stress, the Pearson correlation analysis between ECG characteristics and $CdCl_2$ stress was performed based on the correlation coefficient r and significance p. We first checked the correlation coefficient r to see how much they correlated ($r < 0.3$, poor correlation, $0.3 < r < 0.5$ moderate, $r > 0.5$ high correlation).

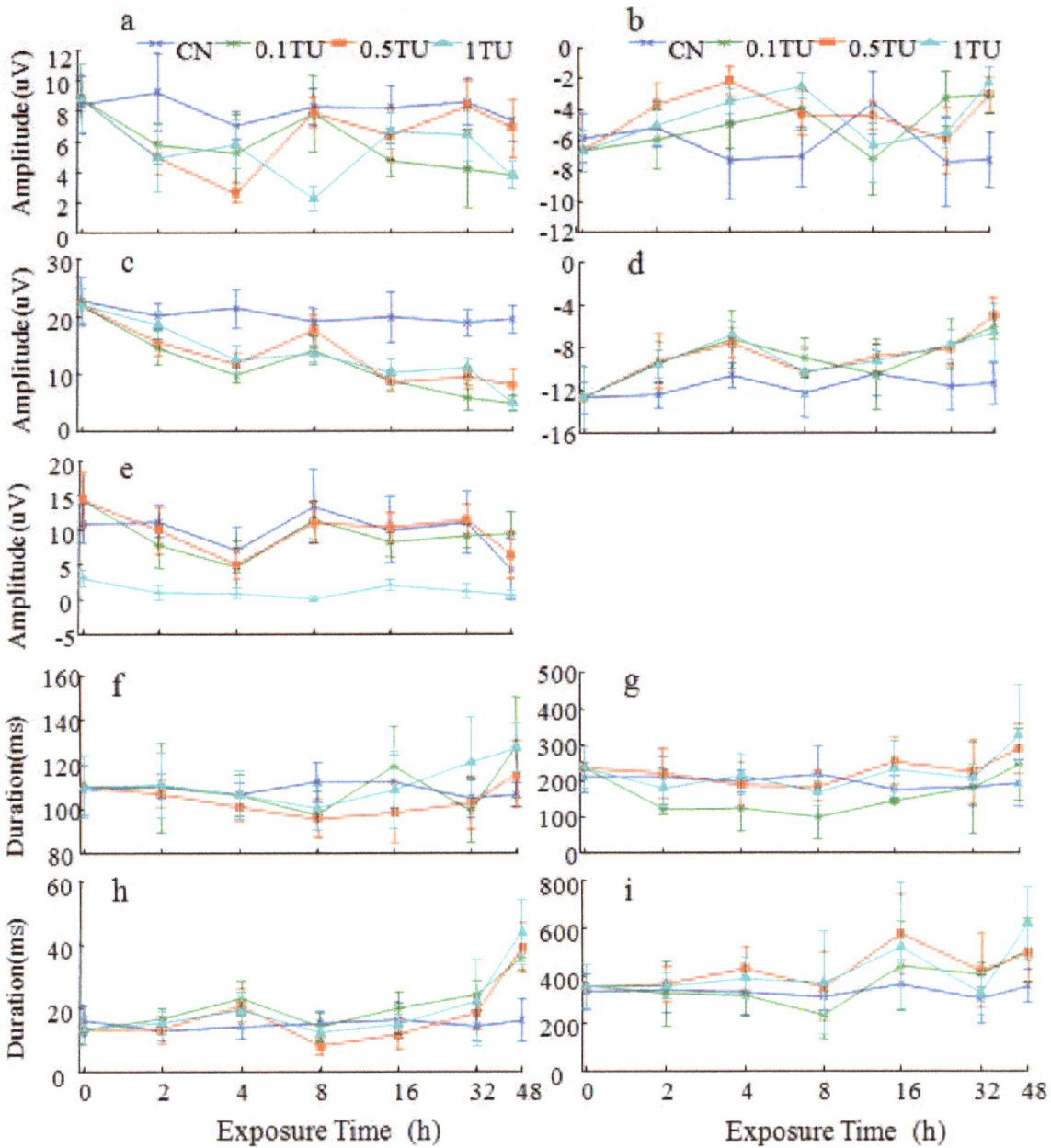

Figure 10. The continuous changes of ECG characteristics in different treatments. a-e represent the characteristics of waves P, Q, R, S and T and f-i represent the characteristics PR interval, QRS complex, of ST segment and QT interval. Log scale was used in the x-axis.

Then, we checked p value to see whether these two variables are correlated significantly ($p < 0.05$). When r is high (absolute $r > 0.5$) with significance ($p < 0.05$), it indicates that the data correlation is significant [53, 54].

The environmental stress E is primarily decided by both chemical concentrations (C) and exposure time (t) with an exponential function as shown in Eq. (2) [55]:

$$E = e^{(C \times t)} + E_f \tag{2}$$

ECG characteristics	Treatments	Start time 0 h	Exposure time 2 h	4 h	8 h	16 h	32 h	48 h
P (μv)	Control	—	\	\	\	\	\	\
	0.1 TU	—	**	**	\	**	**	**
	0.5 TU	—	\	\	\	\	\	\
	1.0 TU	—	**	**	**	*	*	**
Q (μv)	Control	—	\	\	\	**	*	\
	0.1 TU	—	\	*	**	\	**	**
	0.5 TU	—	**	**	**	**	\	**
	1.0 TU	—	*	**	**	**	\	**
R (μv)	Control	—	*	\	**	**	**	**
	0.1 TU	—	**	**	**	**	**	**
	0.5 TU	—	**	**	**	**	**	**
	1.0 TU	—	*	**	**	**	**	**
S (μv)	Control	—	\	*	\	*	\	\
	0.1 TU	—	**	**	**	*	**	**
	0.5 TU	—	**	**	*	**	**	**
	1.0 TU	—	**	**	*	**	**	**
T (μv)	Control	—	\	\	\	\	\	*
	0.1 TU	—	**	**	**	**	**	**
	0.5 TU	—	**	**	*	**	*	**
	1.0 TU	—	**	**	**	*	**	**
PR (ms)	Control	—	\	\	\	\	\	\
	0.1 TU	—	\	\	*	\	\	**
	0.5 TU	—	\	\	*	*	\	\
	1.0 TU	—	\	\	\	\	\	**
QRS (ms)	Control	—	\	\	\	\	\	\
	0.1 TU	—	**	**	**	**	\	\
	0.5 TU	—	\	\	\	\	\	\
	1.0 TU	—	\	\	*	\	\	**
ST (ms)	Control	—	\	\	\	\	\	\
	0.1 TU	—	\	**	\	**	**	**
	0.5 TU	—	\	**	*	\	*	**
	1.0 TU	—	\	*	\	\	**	**

ECG characteristics	Treatments	Start time	Exposure time					
		0 h	2 h	4 h	8 h	16 h	32 h	48 h
QT (ms)	Control	−	\	\	\	\	\	\
	0.1 TU	−	\	\	*	\	\	**
	0.5 TU	−	\	\	\	**	\	**
	1.0 TU	−	\	\	\	**	\	**

The significance shows the difference between exposure times and start time (0 h) in each treatment.* $p < 0.05$, ** $p < 0.01$ and \ represents no significant differences.

Table 2. The difference of zebrafish ECG characteristics after the two-way ANOVA.

in which C is based on TU values and t presents the time with 6 min per unit (0–480) according to our previous results [13]. E_f is the environmental stress due to the effects of all other physico-chemical factors, including water temperature, turbidity, pH, dioxide oxygen and conductivity. As physico-chemical factors are controlled under experimental conditions as shown in our previous research results [56], the effects of E_f are not considered ($E_f = 0$) for simplicity of model execution in this study. Then, if $C \times t = 0$, $E = 1$, it means that the minimum value of E is 1 in a pollution-free environment [57].

The correlation between ECG parameters and $CdCl_2$ stress (E) after Pearson correlation analysis is shown in **Table 3**. The results suggest that the relationship between E and wave S, intervals PR, QRS, ST and QT, showed a high correlation with absolute $r > 0.5$ and $p < 0.05$. With the correlation coefficient $r = 0.729$ (the highest) and correlation significance $p = 0.002$ (the smallest), the relationship between E and QRS was extremely significant, which suggested that QRS could be significantly affected by $CdCl_2$ stress.

On the surface ECG, QRS reflects ventricular depolarization and propagation of the excitatory cardiac impulse throughout the ventricles [58]. Cardiac conduction system excitability was depressed preferentially in Cd^{2+} [59]. Cd^{2+} could lead to changes in ECG, which may be attributed to its effects on ventricular conduction. Yin et al. found that workers exposed to Cd^{2+} had a significantly longer QRS [60]. However, Cd^{2+} administration caused a reduction in myocardiac contractile performance, slowing of heart rate and disturbances in metabolism

Pearson correlation analysis		ECG parameters								
		P	Q	R	S	T	PR	QRS	ST	QT
E	Correlation (r)	−0.094	0.317	−0.487	0.517	−0.443	0.685	**0.729**	0.667	0.635
	Significance (p)	0.739	0.25	0.066	0.048*	0.098	0.005**	**0.002****	0.007**	0.011*

*$p < 0.05$
**$p < 0.01$.

Table 3. The correlation between ECG parameters and environmental stress (E) after Pearson correlation analysis.

of the heart [61], which may prolong QRS. On a cellular level, depolarizing inward current passing through the voltage-gated cardiac sodium (Na^+) channel is responsible for the rapid upstroke of the ventricular action potential that initiates the conduction of the excitatory wave front throughout the ventricular wall [58]. Visentin et al. investigated the dependence of Na^+ current block by Cd^{2+} on external Na^+ concentration in ventricular myocytes. Depression of inward Na^+ current by Cd^{2+} was essentially voltage independent, in agreement with it being caused by channel block. The data show that Cd^{2+} reduces Na^+ current in Purkinje fibres and in ventricular myocytes [62]. In other words, exposure to Cd^{2+} could have an effect on Na + current in zebrafish cardiac myocytes. Cd^{2+} could block the Na^+ channel and decrease inward Na^+ currents, resulting in delayed ventricular conduction and prolonged QRS in the ECG.

4. Conclusions

As a by-product of industry, wastewater with Cd^{2+} should not be discharged into aquatic environment only after basic treatment according to the Environmental Quality Standards for Surface Water, GB3838–2002 [63], in which the limitation of Cd is 0.01 mg/L. However, some industrial wastewater may be discharged without any treatment, in which Cd^{2+} concentration might be higher than 26 mg/L [64] and the concentration of Cd^{2+} in some sediments could reach 359.8 g/kg [65].

The online behaviour responses of zebrafish showed that BS has obvious dose-effect relationship with Cd^{2+}, and the online behaviour responses could illustrate the toxicity of Cd^{2+} directly. The circadian rhythms could be observed even in higher concentrations (1.0 and 2.0 TU). Meanwhile, Cd^{2+} has an inhibitory effect on the standard metabolic rate of zebrafish, and respiratory parameter of zebrafish with ECG can be regarded as sensitive biological monitoring indexes to realize the online assessment of Cd^{2+} pollution. It is noteworthy that there is an extreme significant correlation between QRS complex and Cd^{2+} stress with the highest r and the smallest p among all ECG characteristics, and it may be a good way to monitor Cd^{2+} pollution in aquatic environment by observing and analysing QRS complex.

These results provide an objective ground for analysing complex stress response that could be applied to test the changes of organisms (exposed in different treatments) quantitatively in toxic physiology and ecology.

Acknowledgements

This study was financially supported by the Natural Science Foundation of China (21107135, 41501538), the High-level Talent Project of Shandong Normal University, the Taishan Leader Talent Project of Shandong (tscy20150707), the Oversea High-level Talent Project of Ji'nan (2013041), Jinan Livelihood Major Projects (201704139), Guangdong Provincial Department of Science and Technology (2016A020221016, 2015A020216011), Guangzhou Science Technology and Innovation Commission (201510010294), and Central Research Institutes of Basic Research and Public Service Special Operations (PM-zx703-201701-007).

Author details

Zongming Ren[1] and Yuedan Liu[2]*

*Address all correspondence to: liuyuedan@scies.org

1 Institute of Environment and Ecology, Shandong Normal University, Jinan, PR China

2 The Key Laboratory of Water and Air Pollution Control of Guangdong Province, South China Institute of Environmental Sciences, MEP, Guangzhou, PR China

References

[1] Wang W, Dong C, Dong W, Yang C, Ju T, Huang L, Ren Z. The design and implementation of risk assessment model for hazard installations based on AHP-FCE method: A case study of Nansi Lake Basin. Ecological Informatics. 2016;**36**:162-171

[2] Chau KW. Persistent organic pollution characterization of sediments in Pearl River estuary. Chemosphere. 2006;**64**(9):1545-1549

[3] Yang C, Zhang M, Dong W, Cui G, Ren Z, Wang W. Highly efficient photocatalytic degradation of methylene blue by PoPD/TiO$_2$ nanocomposite. PLoS One. 2017;**12**(3):e0174104

[4] Schwaiger J, Ferling H, Mallow U, Wintermayr H, Negele RD. Toxic effects of the non-steroidal anti-inflammatory drug diclofenac: Part I: Histopathological alterations and bioaccumulation in rainbow trout. Aquatic Toxicology. 2004;**68**(2):141-150

[5] Zhang Y, Wang W, Song J, Ren Z, Yuan H, Yan H, Zhang J, Pei Z, He Z. Environmental characteristics of polybrominated diphenyl ethers (PBDEs) in marine system, with emphasis on marine organisms and sediments. BioMed Research International. 2016b;**20**:1-16

[6] Ohaga SO, Ndiege IO, Kubasu SS, Beier JC, Mbogo CM. Susceptibility of non-target aquatic macro-invertebrates and vertebrates to (*piperaceae*) and (*asteraceae*) powder in Kilifi district, Kenya. International Journal of Zoological Research. 2007;**3**(2):86-93

[7] Ren Z, Chon T-S, Xia C, Li F. The monitoring and assessment of aquatic toxicology. BioMed Research International. 2017;**2017**:9179728

[8] Parsons M, Thoms M, Norris R. Australian river assessment system: AusRivAS physical assessment protocol. In: Monitoring River Health Initiative Technical Report. 22. 2002

[9] Zhang D, Ding A, Lin X, Sun Y, Zheng L, Chen C. Monitoring and assessment of water quality in rivers with biological methods. Journal of Beijing Normal University (Natural Science). 2009;**45**(2):200-204. (in Chinese)

[10] Ren Z, Li Z, Zha J, Ma M, Wang Z, Fu R. Behavioral responses of Daphnia magna to stresses of contaminants with different toxic mechanisms. Environmental Contamination and Toxicology. 2009;**82**(3):310-316

[11] Campanella L, Cubadda F, Sammartino MP, Saoncella A. An algal biosensor for the monitoring of water toxicity in estuarine environments. Water Research. 2001;**35**(1):69

[12] Glasgow HB, Burkholder JAM, Reed RE, Lewitus AJ, Kleinman JE. Real-time remote monitoring of water quality: a review of current applications and advancements in sensor, telemetry and computing technologies. Journal of Experimental Marine Biology & Ecology. 2004;**300**(1-2):409-448

[13] Zhang G, Chen L, Chen J, Ren Z, Wang Z, Chon T-S. Evidence for the Stepwise Behavioral Response Model (SBRM): The effects of carbamate pesticides on medaka (*Oryzias latipes*) in an online monitoring system. Chemosphere. 2012;**87**:734-741

[14] Alexakis D. Assessment of water quality in the messolonghi-etoliko and neochorio region (west greece) using hydrochemical and statistical analysis methods. Environmental Monitoring & Assessment. 2011;**182**(1-4):397-413

[15] Allen P. Mercury accumulation profiles and their modification by interaction with cadmium and lead in the soft tissues of the cichlid *Oreochromis aureus* during chronic exposure. Bulletin of Environmental Contamination and Toxicology. 1994;**53**(5):684-692

[16] Hansen LG, Lambert RJ. Transfer of toxic trace substances by way of food animals: Selected examples. [felis catus; sus scrofa domesticus; rattus norvegicus]. Journal of Environmental Quality. 2010;**16**:3

[17] Suzuki Y, Nakamura R, Ueda T. Accumulation of strontium and calcium in freshwater fishes of japan. Journal of Radiation Research. 1972;**13**(4):199-207

[18] Zhang X, Ren B, Li S, Qu X, Yang H, Xu S, Ren Z, Kong Q, Wang C, Chon T-S. Is sodium percarbonate a good choice in online degrading Deltamethrin? Frontiers of Environmental Science & Engineering. 2017;**11**(3):3

[19] Ren Z, Li Z, Zha J, Ma M, Wang Z, Gerhardt A. The early warning of aquatic organophosphorus pesticide contamination by on-line monitoring behavioral changes of Daphnia magna. Environmental Monitoring and Assessment. 2007;**134**(1-3):373-383

[20] Ren Z, Liu L, Fu R, Miao M. The stepwise behavioral responses: Behavioral adjustment of the Chinese rare minnow (*Gobiocypris rarus*) in the exposure of carbamate pesticides. BioMed Research International. 2013;**2013**:697279-697287

[21] Colgan P. Motivational analysis of fish feeding. Behaviour. 1973;**45**(1-2):38-65

[22] Li Z, Ren Z, Ma M, Zha J, Wang Z, Fu R. Research on the behavioral responses of *Daphnia magna* in the early warning of organophosphorous pesticide contamination. China Water and Wastewater. 2007;**23**(12):73-75 (In Chinese)

[23] Selye H. The evolution of the stress concept. American Scientist. 1973;**61**:692-699

[24] Ren Z, Li Z, Zha J, Rao K, Ma M, Wang Z, Fu R. The avoidance responses of *Daphnia magna* to the exposure of organophosphorus pesticides in an on-line biomonitoring system. Environmental Modeling and Assessment. 2009;**14**(3):405-410

[25] Ren Z, Zhang X, Wang X, Qi P, Zhang B, Zeng Y, Fu R, Miao M. AChE inhibition: One dominant factor for swimming behavior changes of *Daphnia magna* under DDVP exposure. Chemosphere. 2015;**120**:252-257

[26] Wang L, Ren Z, Kim H, Xia C, Fu R, Chon T-S. Characterizing response behavior of medaka (*Oryzias latipes*) under chemical stress based on self-organizing map and filtering by integration. Ecological Informatics. 2015;**29**:107-118

[27] Brown JH, Gillooly JF, Allen AP, Savage VM, West GB. Response to forum commentary on "toward a metabolic theory of ecology". Ecology. 2004;**85**(7):1818-1821

[28] Qi L, Song J, Li S, Cui X, Peng X, Wang W, Ren Z, Han M, Zhang Y. The physiological characteristics of zebra fish (*Danio rerio*) based on metabolism and behavior: a new method for the online assessment of cadmium stress. Chemosphere. 2017;**184**:1150-1156

[29] Vichkovitten T, Inoue H. Dissolved oxygen environments of the fish farm in Yashima Bay, Japan I. Primary productivity of sea water. Kasetsart Journal (Natural Science). 1997;**31**:385-391

[30] Goldberger AL. Clinical electrocardiography: A simplified approach. 9 th ed. 2012;**347**:17-39

[31] Einthoven W. The different forms of the human electrocardiogram and their signification. Lancet. 1912;**179**(4622):853-861

[32] Berul CI, Aronovitz MJ, Wang PJ, Mendelsohn ME. In vivo cardiac electrophysiology studies in the mouse. Circulation. 1996;**94**(10):2641-2648

[33] Machida N, Aohagi Y. Electrocardiography, heart rates and heart weights of free-living birds. Journal of Zoo & Wildlife Medicin. 2001;**32**(32):47-54

[34] Xing N, Ji L, Song j, Ma J, Li J, Ren Z, Xu F, Zhu J. Cadmium stress assessment based on the electrocardiogram characteristics of zebra fish (*Danio rerio*): QRS complex could play animportant role. Aquatic Toxicology. 2017;**191**:236-244

[35] Langheinrich U, Vacun G, Wagner T. Zebrafish embryos express an orthologue of HERG and are sensitive toward a range of QT-prolonging drugs inducing severe arrhythmia☆[J]. Toxicology and Applied Pharmacology. 2003;**193**(3):370-382

[36] Zhang X, Yang H, Ren Z, Cui Z. The toxic effects of deltamethrin on Zebrafish: the correlation among behavior response, physiological damage and AChE. RSC Advances. 2016a;**6**(111)

[37] Chen SL, Korzh V, Strahle U. Zebrafish embryos are susceptible to the dopaminergic neurotoxin mptp. European Journal of Neuroscience. 2005;**21**(6):1758-1762

[38] Zhang T, Yang M, Pan H, Li S, Ren B, Ren Z, Xing N, Qi L, Ren Q, Xu S, Song J, Ma J. Does time difference of the acetylcholinesterase (ache) inhibition in different tissues exist? A case study of zebra fish (zebrafish) exposed to cadmium chloride and deltamethrin. Chemosphere. 2017;**168**:908-916

[39] Petersen GI. Suitability of Zebrafish as Test Organism for Detection of Endocrine Disrupting Chemicals. Nordic Council of Ministers. Copenhagen, Denmark. 2001

[40] Ren Z, Wang Z. The differences in the behavior characteristics between *Daphnia magna* and Japanese Madaka in an on-line biomonitoring system. Journal of Environmental Science. 2010;**22**(5):703-708

[41] OECD. Test No. 203: Fish, acute toxicity test. OECD Guidelines for the Testing of Chemicals. 2006;**2**:1-10

[42] Holford NHG, Sheiner LB. Understanding the dose-effect relationship. Clinical Pharmacokinetics. 1981;**6**(6):429-453

[43] Hamblen-Coyle MJ, Wheeler DA, Rutila JE, Rosbash M, Hall JC. Behavior of period-altered circadian rhythm mutants of Drosophila in light: Dark cycles (Diptera: Drosophilidae). Journal of Insect Behavior. 1992;**5**(4):417-446

[44] Nimkerdphol K, Nakagawa M. Effect of sodium hypochlorite on zebrafish swimming behavior estimated by fractal dimension analysis. Journal of Bioscience and Bioengineering. 2008;**105**(5):486-492

[45] Steinach M, Gunga HC. Circadian Rhythm and Stress. In: Choukér A, editor. Stress Challenges and Immunity in Space: From Mechanisms to Monitoring and Preventive Strategies. Berlin, Heidelberg: Springer Verlag; 2012

[46] Reddy SLN, Venugopal NBRK. Effect of cadmium on acetylcholinesterase activity and oxygen consumption in a freshwater field crab, *Barytelphusa guerini*. Journal of Environmental Biology. 1993;**14**(3):203-210

[47] Bayne BL et al. The Effects of Stress and Pollution on Marine Animals. Praeger; New York. 1985

[48] Siddiqui AA. Respiratory response and hemolymph sugar level of the crab, Barytelphusa gureini exposed to cadmium chloride. International Journal of Research in Engineering and Applied Sciences. 2015;**5**(3):71-78

[49] Zimarino M, Montebello E, Radico F, Gallina S, Perfetti M, Iachini Bellisarii F, De Caterina R. ST segment/heart rate hysteresis improves the diagnostic accuracy of ECG stress test for coronary artery disease in patients with left ventricular hypertrophy. European Journal of Preventive Cardiology. 2016;**23**(15):1632-1639

[50] Mohammed HE, Ahmed SH, Nabil ZI. Electrophysiological protective and therapeutic efficacy of resveratrol against cadmium chloride: An in vitro study. In: Proceedings of the 7th International Conference on Biological Sciences (Zool). 2012. pp. 24-37

[51] Fermini B, Fossa AA. The impact of drug-induced qt interval prolongation on drug discovery and development. Nature Reviews Drug Discovery. 2003;**2**(6):439-447

[52] Jensen RA, Acton EM, Peters JH. Doxorubicin cardiotoxicity in the rat: comparison of electrocardiogram, transmembrane potential and structural effects. Journal of Cardiovascular Pharmacology. 1984;**6**(1):186-200

[53] Chen J, Mehta RS, Baek HM, Nie K, Liu H, Lin M, Yu H, Nalcioglu O, Su M. Clinical characteristics and biomarkers of breast cancer associated with choline concentration measured by 1 h mrs. NMR in Biomedicine. 2011;**24**(3):316-324

[54] Ren Q, Zhang T, Li S, Ren Z, Yang M, Pan H, Xu S, Qi L, Chon T-S. Integrative character-ization of toxic response of zebra fish (Zebrafish) to deltamethrin based on AChE activ-ity and behavior strength. BioMed Research International. 2016;**2016**:7309184

[55] Ren Z, Li S, Zhang T, Qi L, Xing N, Yu H, Jian J, Chon T-S, Bo T. Behavior persistence in defining threshold switch in stepwise response of aquatic organisms exposed to toxic chemicals. Chemosphere. 2016;**165**:409-417

[56] Chen L, Fu X, Zhang G, Zeng Y, Ren Z. Influences of temperature, pH and turbidity on the behavioral responses of *Daphnia magna* and Medaka (*Oryzias latipes*) in the biomoni-tor. Procedia Environmental Sciences. 2012;**13**:80-86

[57] Yin L, Yang H, Si G, Ren Q, Fu R, Zhang B, Zhang X, Wang X, Qi P, Xia C, Ren Z, Chon T-S, Kim H. Persistence parameter: A reliable measurement for behavioral responses of medaka (*Oryzias latipes*) to environmental stress. Environmental Modeling and Assess-ment. 2016;**21**(1):159-167

[58] Harmer AR, Valentin JP, Pollard CE. On the relationship between block of the cardiac Na+ channel and drug-induced prolongation of the qrs complex. British Journal of Pharmacology. 2011;**164**(2):260

[59] Kopp SJ, Perry HM Jr, Perry EF, Erlanger M. Cardiac physiologic and tissue metabolic changes following chronic low-level cadmium and cadmium plus lead ingestion in the rat. Toxicology & Applied Pharmacology. 1983;**69**(1):149-160

[60] Yin ZY, Jin TY, Yun Ping HU, Hong F. Cardiovascular function in workers exposed to cadmium. Journal of Labour Medicine. 2003. DOI: 10.3969/j.issn.1006-3617.2003.03.011

[61] Kisling GM, Kopp SJ, Paulson DJ, Tow JP. Cadmium-induced attenuation of coronary blood flow in the perfused rat heart. Toxicology & Applied Pharmacology. 1993;**118**(1): 58-64

[62] Visentin S, Zaza A, Ferroni A, Tromba C, Difrancesco C. Sodium current block caused by group iib cations in calf purkinje fibres and in guinea-pig ventricular myocytes. Pflügers Archiv European Journal of Physiology. 1990;**417**(2):213-222

[63] GB3838-2002. The Surface Water Environmental Quality Standard. China Environmental Science Press; 2002 (in Chinese)

[64] He H, Zhu T, Liu Z, Chen M, Wang L. Treatment of cadmium-containing wastewater in zinc smelting using sodium dimethyl dithio carbamate. Environmental Protection of Chemical Industry. 2015;**35**(3):293-296 (in Chinese)

[65] Huang Z, Zhou Z, Luo Y. Distribution of Cd in sediments from Xiawan drainage area of the Xiang River. Environment Pollution and Protection. 2009;**31**(7):56-58 (in Chinese)

3

Zebrafish or *Danio rerio*: A New Model in Nanotoxicology Study

Maria Violetta Brundo and Antonio Salvaggio

Abstract

Nanotoxicology represents a new research area in toxicology that allows to evaluate the toxicological properties of nanoparticles in order to determine whether and to what extent they represent an environmental threat. Behavior, fate, transport, and toxicity of nanoparticles are influenced to their particular properties and of several environmental factors. The mechanisms underlying the toxicity of nanomaterials have recently been studied specially in aquatic organisms. In particular, in recent years, the use of *Danio rerio* or zebrafish as an animal model system for nanoparticle toxicity assay increased exponentially. In this review, we compare the recent researches employing zebrafish, adults or embryos, for different nanoparticles' toxicity assessment.

Keywords: *Danio rerio*, nanoparticles, nanomaterials, biomarkers, ZFET

1. Introduction

Nanotechnology has advanced exponentially over the past decade, and nanoscale materials are being exploited in several applications [1]. Between 2011 and 2015, there has been a 30-fold increase in the production of nanoproducts [2].

Engineered nanodevices are finding a new range of applications for the possibility of modifications of their shape, size, surface, and chemical properties. These characteristics are not present in their bulk counterparts [1]. For example, they have very high specific surface areas that give rise to enhanced reactivity and solubility, reduced melting and sintering temperatures, as well as altered crystal structures [1].

Nowadays, we are using a wide variety of commercially available nanoparticles [1]. Metal and carbon nanoparticles (NPs) represent the largest and fastest growing group of NPs [2].

Hence, environmental contamination is already occurring and is predicted to increase dramatically. This growth of nanotechnology has not advanced without concerns regarding their potential adverse environmental impacts. Several studies of nanotoxicology have been made in fact to evaluate the toxicity of various NPs [3–5]. However, there is much to do for evaluating whether the NPs may be an environmental threat. In fact, there are an increasing number of studies where the toxicity of a several engineered nanoparticles or nanomaterials such as fullerenes, graphene metal nanoparticles, metal oxides nanoparticles, crystalline materials, amorphous materials, and nano-sized polymers has been evaluated [6]. For toxicity assays several model organisms are used, such as *Daphnia magna* [7–9], *Paracentrotus lividus* [10, 11], *Mytilus galloprovincialis* [12, 13] or *Mytilus edulis* [14, 15], *Artemia salina* [16, 17], *Danio rerio* [18–21], and other animal models like mice [22, 23]. Each and every year, the number of engineered nanomaterials and their products is continuously increasing, which is necessary to have a representative model organism, able to assess nanotoxicity accurately. In this regard, zebrafish as model organism has attracted scientific interest. In this review, we focused on nanotoxicological studies conducted using as model zebrafish, adults or embryos, with different assessment methods

2. Zebrafish: a perfect experimental animal model

Danio rerio (Hamilton and Buchanan, 1822), commonly known as zebrafish, is a small tropical freshwater fish which lives in river basins of India, Northern Pakistan, Nepal, Bhutan, and South Asia. It belongs to the family of Cyprinidae, within the order of the Cypriniformes. Zebrafish adults are 4–5 cm long, and so they can be easily managed in large numbers in the laboratory.

Zebrafish can tolerate a temperature range of 24.5–28.5°C [24]; however, the growth speed of zebrafish embryos varies according to temperature [25].

One of the reasons why zebrafish is an excellent laboratory model is for its ability to spawn huge amounts of eggs the whole year.

Zebrafish embryonic development has been well characterized to [25]. The embryos themselves are transparent during the first few days of their lives because chorion is transparent. Pigmentation in the embryos starts about 30–72 h post fertilization [26]. Fertilization activates cytoplasmic movements, easily evident within about 10 min. The first cleavage of the newly fertilized egg occurs about 40 min after fertilization. The cytoplasmic divisions are meroblastic, and at the end of them, a blastodisc forms.

Blastula of zebrafish is a "stereoblastula" because blastocoel is not present. The blastula and gastrula stage of zebrafish at 28°C is equivalent to 2.25–5.25 h post fertilization (hpf) and 5.25–10 hpf, respectively [25].

Ballard [27] coined the term "pharyngula" (24–48 h) to refer to the embryo that has developed to the phylotypic stage, when it possesses the classic vertebrate bauplan. This is the time of development when one can most readily compare the morphologies of embryos of diverse vertebrates.

During hatching period (48–72 h), they are called "embryos" until the end of the third day and afterward "larvae," whether they have hatched or not [25].

There are several advantages for using zebrafish as a model species in nanotoxicological studies. Main benefit regards its size. Zebrafish adult is approximately 5 cm long, so it can be handled without any difficulty and reduces housing space and husbandry costs.

The tiny size of the larval and adult zebrafish allows to reduce quantities the dosing of experimental solutions and thereby creates limited volumes of waste to disposal and minimizes quantities of lab ware and chemicals.

Small embryos allow reasonable sample sizes to be tested together using a multiwell plate or series of Petri dishes to provide several experimental replicates at one time. From the egg stage, zebrafish embryos can survive for several days through the absorption of yolk and can be visually assessed for malformation [26].

The rapid maturation of zebrafish (sexual maturation occurs around 100 days) allows easy experimentation for transgenerational endpoints required for mutagenesis screening and assessing chemicals for teratogenicity.

This species shows high fecundity (single female can lay up to 200 eggs per week) and transparent embryos. The eggs hatch rapidly and organogenesis occurs quickly. As a result, the major organs are developed within few days post fertilization (dpf) in larvae.

Zebrafish eggs remain transparent from fertilization to when the tissues become dense and pigmentation is initiated (at approximately 30–72 h post fertilization (hpf)); this allows unobstructed observations of the main morphological changes up to and beyond pharyngulation. Therefore, using little magnification, adverse effects of chemical exposure on development of the brain, notochord, heart, jaw, trunk segmentation, and measurements of size can be assessed quantitatively.

Their optical clarity allows for identification of phenotypic traits during mutagenesis screening and assessment of endpoints of toxicity during toxicity testing. This proves even more valuable when immunochemistry (IHC) techniques are used. There are a vast amount of immunohistochemical markers available, allowing assessments of aberrant morphology or activation of certain signaling pathways by toxicants through the staining of specific tissues and cells types.

The zebrafish research community has developed a range of resources very useful to the toxicologists, including mutant strains, cDNA clone collections, and whole genome that has been sequenced a few years ago. Highly conserved signaling pathways are found both in zebrafish and mammals with a high level of genomic homology [28].

In recent years, the use of zebrafish as an established animal model system for nanoparticle toxicity assay is growing exponentially. Different types of parameters are used to evaluate nanoparticle toxicity such as hatching achievement rate, developmental malformation of organs, damage in gills and skin, abnormal behavior (movement impairment), reproduction toxicity, and finally mortality. In fact, there are an increasing number of literatures that document the concern over toxicity for broad range of engineered nanoparticles or nanomaterials. In this regard, zebrafish as an in vivo model organism has attracted scientific interest because of its unique features abovementioned.

Interestingly, zebrafish behavioral response is also a sensitive indicator for abnormal change in toxicity. An experiment performed by [29] has also shown that TiO_2 nanoparticles affect larval swimming parameters, including velocity and activity level.

The disruption of gills, skin, and endocrine system by nanoparticles is another parameter to understand nanoparticle-induced toxicity. It has been reported that silver ions (Ag^+) generated by AgNPs exert acute toxicity, mainly due to their interaction with the gills. In the gills, ionic Ag^+ inhibits Na^+/K^+-ATPase action and the enzymes related to Na^+ and Cl^- uptake, finally affecting osmoregulation [30].

Nanoparticle affects male and female reproductivity and fetal development. Wang et al. [31] assessed the disturbance in zebrafish reproduction after the chronic exposure of TiO_2 nanoparticles.

Using this model organism, several specific protocols have been used for the toxicity screening. The correlation of successful hatching efficiency and embryo toxicity is an important parameter to evaluate the nanotoxicity.

2.1. Hatching analysis

The hatching-related parameters may be one of the endpoints that have been underestimated in the several studies. There are conflicting results about the endpoint in the same and different species [32, 33], because the results are not easy to interpret. Consequently, hatching-related parameters do not seem to be able to show the toxicity of a nanoparticle especially at environmental-relevant concentrations. In fact, many papers associate different endpoints related to hatching and embryo development [34].

Paatero et al. [35] have used *Danio rerio* embryos to study toxicity profiles of differently surface-functionalized mesoporous silica nanoparticles (MSNs). Embryos with the chorion membrane intact, or dechorionated embryos, were incubated or microinjected with amino (NH2-MSNs), polyethyleneimine (PEI-MSNs), succinic acid (SUCC-MSNs), or polyethylene glycol (PEG-MSNs)-functionalized MSNs. Toxicity was assessed by viability and cardiovascular function. Typically cardiovascular toxicity was evident prior to lethality. Confocal microscopy revealed that PEI-MSNs penetrated into the embryos, whereas PEG^-, NH_2^-, and SUCC-MSNs remained aggregated on the skin surface. Direct exposure of inner organs by microinjecting NH_2-MSNs and PEI-MSNs demonstrated that the particles displayed similar toxicity indicating that functionalization affects the toxicity profile by influencing penetrance through biological barriers.

Samaee et al. [36] have studied the nano-TiO$_2$ toxicity to zebrafish embryos through evaluating the success in hatching in relationship with hours postexposure. Zebrafish embryos 4 h post fertilization were exposed to nTiO$_2$ (0, 0.01, 10, and 1000 µg mL(−1)) for 130 h. The hatching rate (HR) was calculated for each concentration tested. It was observed that TiO$_2$ nanoparticles can cause premature hatching in zebrafish embryos, dose dependently.

Ong et al. [37] have used silicon, cadmium selenide, silver, and zinc NPs as well as single-walled carbon nanotubes to assess NP effects on zebrafish hatch. They have reported complete inhibition of hatching and embryo death within chorion upon nanoparticle exposure, because the nanoparticles interact with the hatching enzymes, and they concluded that the observed effects arose from the NPs themselves and not their dissolved metal components.

2.2. Developmental disorder analysis of zebrafish embryos: zebrafish embryo toxicity test

Zebrafish embryo toxicity test (ZFET) is a modern nonanimal test, and it represents an alternative approach to acute toxicity testing, since with the same sensitivity and specificity, it is possible to find more simplification, economicity, and speedy of execution, as well as suggested by the European Community in order to decrease the impact of the experimental tests on live animals [20, 38]. Fish Embryo Toxicity Test is included in the guidelines to perform toxicity test about FDA and ICH for the pharmaceutical products and about EPA and OECD for the chemical substances [39].

Fish embryo-larval assays provide a screening and investigative tool able of testing a larger number of nanoparticles, and this model has become increasingly common for investigation of developmental toxicity mechanisms [18–20, 40, 41]. ZFET is not a suitable test if you want to evaluate the developmental malformations after the 96 hours post fertilization (hpf), such as the skeletal anomalies, because calcification process in zebrafish starts the seventh day of development.

Usenko et al. [42] have evaluated carbon fullerene (C$_{60}$, C$_{70}$, and C$_{60}$(OH)$_{24}$) toxicity in zebrafish embryos. They observed caudal fin malformation at the concentrations of 200 ppb of C$_{60}$ and C$_{70}$ and yolk sac edema, pericardial edema, and pectoral fin malformations over the concentrations of 2500 ppb of C$_{60}$(OH)$_{24}$. Additionally, they also observed swelling of zebrafish embryos and delay in development upon exposure to 5000 ppb of C$_{60}$(OH)$_{24}$.

Brundo et al. [18] tested the nanomaterials that were synthesized proposing a groundbreaking approach by an upside-down vision of the Au/TiO$_2$ nano-system to avoid the release of nanoparticles. The system was synthesized by wrapping Au nanoparticles with a thin layer of TiO$_2$. The nontoxicity of the nano-system was established by testing the effect of the material on *Danio rerio* larvae. Zebrafish larvae were exposed to different concentrations of nanoparticles of TiO$_2$ and Au and to new nanomaterials. Authors evaluated as biomarkers of exposure the expression of inducible metallothioneins. The results obtained by toxicity test showed that neither mortality nor sublethal effects were induced by the different nanomaterials and free nanoparticles tested. However, only zebrafish larvae exposed to free Au nanoparticles showed a different response to anti-MT antibody. In fact, the immunolocalization analysis highlighted an increase synthesis of the inducible metallothioneins.

Xu et al. [41] evaluated the effect of CuO-NPs on early zebrafish development. The results reveal that CuO-NPs can induce abnormal phenotypes of a smaller head and eyes and delayed epiboly. The gene expression pattern shows that CuO-NPs spatially narrow the expression of dorsal genes chordin and goosecoid and alter the expression of dlx3, ntl, and hgg which are related to the cell migration of gastrulation. The decreased expression of pax2 and pax6 involved in neural differentiation was accordant with the decreased sizes of neural structures. $Cmlc_2$ expression suggests that CuO-NPs prevented looping of the heart tube during cardiogenesis. Furthermore, quantitative RT-PCR results suggest that the CuO-NPs could increase the canonical Wnt signaling pathway to narrow the expression of chordin and goosecoid in dorsoventral patterning as well as decrease the transcription of Wnt5 and Wnt11 to result in slower, less directed movements and an abnormal cell shape. These findings indicated the CuO-NPs exert developmental toxicity.

Pecoraro et al. [20] have tested nanocomposite membranes prepared using Nafion polymer combined with various fillers, such as anatase-type TiO_2 nanoparticles and graphene oxide. The nontoxicity of these nanocomposites, already shown to be effective for water purification applications [19], was recognized by testing the effect of these different materials on zebrafish embryos. They evaluated as biomarkers of exposure the expression of heme-oxygenase 1 and inducible nitric oxide synthases. Embryo toxicity test showed that neither mortality nor sublethal effects were caused by the different nanoparticles and nanosystems tested. Only zebrafish larvae exposed to free nanoparticles have shown a different response to antibodies anti-heme-oxygenase 1 and anti-inducible nitric oxide synthases. The immunolocalization analysis in fact has highlighted an increase in the synthesis of these biomarkers.

2.3. Pathologies analysis in organs of zebrafish embryos and adults

As no authorization is required, many authors prefer the ZFET, and few toxicity studies on nanoparticles are conducted with embryos after 96 hpf. Developmental malformation of zebrafish embryos was studied to several authors, and they can relate incomplete organ development, deformity of body parts, or lack of pigmentation. Zhu et al. [43] did one of the first studies on developmental toxicity in fish caused by iron oxide nanoparticles in aquatic environments. To study the ecological effects of iron oxide nanoparticles on aquatic organisms, they used early life stages of the *Danio rerio* to examine such effects on embryonic development in this species. The results showed that ≥10 mg/L of iron oxide nanoparticles instigated developmental toxicity in these embryos, causing mortality, hatching delay, and malformation. Moreover, an early life stage test using zebrafish embryos/larvae is also discussed and recommended in this study as an effective protocol for assessing the potential toxicity of nanoparticles.

Pecoraro et al. [21] did a study on adverse effects of AgNPs in adult of *Danio rerio*. Fishes exposed to increasing concentrations (8, 45, 70 µg/l) of silver nanoparticles (AgNPs, 25 nm in average diameter) and after treatment for 30 days were quickly euthanized in MS-222. Authors have evaluated bioaccumulation of AgNPs using ICP-MS and analyzed histological changes, biomarkers of oxidative damage, and gene expression in the gut, liver, and gill tissues of AgNPs-treated zebrafish. The histological analysis showed lesions of secondary lamellae of the gills with different degrees of toxicity such as hyperplasia, lamellar fusion, subepithelial edema, and even in some cases telangiectasia. Huge necrosis of the intestinal villi was found in the gut. No lesion was detected in the liver. The analysis revealed a high expression of metallothioneins 1 (MTs 1) in animals exposed to AgNPs compared to the control group. The ICP-MS analysis shows that the amount of particles absorbed in all treated samples is almost the same. They affirm that AgNP toxicity is linked more to their size and state of aggregation than to their concentrations. Silver nanoparticles can damage gills and gut because they are able to pass through the mucosal barrier thanks to their small size. However, the damage is still reversible because it is not documented injury to the basal membrane.

3. Discussion

The toxicology of engineered NMs is a relatively new and evolving field, and although their applications are increasing, there are many concerns about their environmental and health impacts [44]. A large number of studies carried out on several nanoparticles have produced conflicting results. In fact, despite continuous attempts to establish a correlation between structure of the particles and their interactions with biological systems, we are still far from elucidating with certainty the toxicological profile of NPs [45]. Among these investigations, a large numbers of authors, for example, have confirmed the nontoxicity of AuNPs [46–50]; conversely, others have observed the toxicity of AuNPs [51–53].

Despite some authors showed low toxicity of other particles such as TiO_2 NPs [54, 55], studies demonstrated that exposure to high concentrations of TiO_2 particles was able to induce lung tumor formation after 2 years in rats [56]. Moreover, the International Agency for Research on Cancer (IARC) has classified TiO_2 as a possibly carcinogenic to human (Group 2B carcinogen) [57].

For this reason, same researches are developing an innovative nanomaterial that could help to overcome problems related to the toxic effects of NPs, being able to exploit all their qualities [12, 18, 20].

The use of zebrafish as animal model is recommended in several of these researches because it is an inexpensive, quick, and easy model to assess the nanoparticle toxicity [58], and it can offer many advantages for toxicological research [59, 60]. In particular, ZFET, an alternative approach to acute toxicity testing, is important in order to decrease the impact of the experimental tests on live animals as well as suggested by the European Community. Therefore, the use of zebrafish model can be proposed for screening the toxicity profile of nanomaterials and their rapid feedback [61].

Author details

Maria Violetta Brundo[1]* and Antonio Salvaggio[2]

*Address all correspondence to: mvbrundo@unict.it

1 Department of Biological, Geological and Environmental Science, University of Catania, Catania, Italy

2 Experimental Zooprophylactic Institute of Sicily "A. Mirri", Catania, Italy

References

[1] Tsuzuki T. Commercial-scale Production of Nanoparticles. Boca Raton: CRC Press, Taylor & Francis Group; 2013. pp. 978-981

[2] Vance ME, Kuiken T, Vejerano EP, McGinnis SP, Hochelle MF Jr, Rejeski D, Hull MS. Nanotechnology in the real world: Redeveloping the nanomaterials consumer products inventory. Journal of Nanotechnology. 2015;**6**:1769-1780. DOI: 10.3762/bjnano.6.181

[3] Oberdorster E, Zhu SQ, Blickley TM, Clellan-Green P, Aasch ML. Ecotoxicology of carbon-based engineered nanoparticles: Effects of fullerene (C-60) on aquatic organisms. Carbon. 2006;**44**(6):1112-1120. DOI: 10.1016/j.carbon.2005.11.008

[4] Moore MN. Do nanoparticles present ecotoxicological risks for the health of the aquatic environment? Environment International. 2006;**32**(8):967-976. DOI: 10.1016/j.envint.2006.06.014

[5] Chakraborty C, Sharma AR, Sharma G, Lee SS. Zebrafish: A complete animal model to enumerate the nanoparticle toxicity. Journal of Nanbiotechnology. 2016;**14**:65. DOI: 10.1186/s12951-016-0217-6

[6] Seaton A, Tran L, Aitken R, Donaldson K. Nanoparticles, human health hazard and regulation. Journal of the Royal Society Interface. 2010;**7**(1):S119-S129

[7] Xiao Y, Vijver MG, Chen G, Peijnenburg WJ. Toxicity and accumulation of Cu and ZnO nanoparticles in *Daphnia magna*. Environmental Science & Technology. 2015;**49**(7): 4657-4664. DOI: 10.1021/acs.est.5b00538

[8] Xiao Y, Peijnenburg WJ, Chen G, Vijver MG. Toxicity of copper nanoparticles to *Daphnia magna* under different exposure conditions. Science of the Total Environment. 2016;**1**: 563-564 (81-88). DOI: 10.1016/j.scitotenv.2016.04.104

[9] Pakrashi S, Tan C, Wang WX. Bioaccumulation-based silver nanoparticle toxicity in *Daphnia magna* and maternal impacts. Environmental Toxicology and Chemistry. 2017;**36**(12):3359-3366. DOI: 10.1002/etc.3917

[10] Šiller L, Lemloh ML, Piticharoenphun S, Mendis BG, Horrocks BR, Brümmer F, Medaković D. Silver nanoparticle toxicity in sea urchin *Paracentrotus lividus*. Environmental Pollution. 2013;**178**:498-502. DOI: doi.org/10.1016/j.envpol.2013.03.010

[11] Kanold JM, Wang J, Brümmer F, Šiller L. Metallic nickel nanoparticles and their effect on the embryonic development of the sea urchin *Paracentrotus lividus*. Environmental Pollution. 2016;**212**:224-229. DOI: doi.org/10.1016/j.envpol.2016.01.050

[12] Scuderi V, Impellizzeri G, Romano L, Scuderi M, Brundo MV, Bergum K, Zimbone M, Sanz R, Buccheri MA, Simone F, Nicotra G, Svensson BG, Grimaldi MG, Privitera V. An enhanced photocatalytic response of nanometric TiO_2 wrapping of Au nanoparticles for eco-friendly water applications. Nanoscale. 2014;**6**:11189-11195. DOI: 10.1039/c4nr02820a

[13] Gornati R, Longo A, Rossi F, Maisano M, Sabatino G, Mauceri A, Bernardini G, Fasulo S. Effects of titanium dioxide nanoparticle exposure in *Mytilus galloprovincialis* gills and digestive gland. Nanotoxicology. 2016;**10**(6):807-817. DOI: 10.3109/17435390.2015.1132348

[14] Tedesco S, Doyle H, Iacopino D, O'Donovan I, Keane S, Sheehan D. Gold nanoparticles and oxidative stress in the blue mussel, *Mytilus edulis*. Methods in Molecular Biology. 2013;**1028**:197-203. DOI: 10.1007/978-1-62703-475-3_12

[15] Hu W, Culloty S, Darmody G, Lynch S, Davenport J, Ramirez-Garcia S, Dawson KA, Lynch I, Blasco J, Sheehan D. Toxicity of copper oxide nanoparticles in the blue mussel, *Mytilus edulis*: A redox proteomic investigation. Chemosphere. 2014;**108**:289-299. DOI: 10.1016/j.chemosphere.2014.01.054

[16] Arulvasu C, Jennifer SM, Prabhu D, Chandhirasekar D. Toxicity effect of silver nanoparticles in brine shrimp *Artemia*. Scientific World Journal. 2014;**2014**:10. DOI: 10.1155/2014/256919

[17] Wang C, Jia H, Zhu L, Zhang H, Wang Y. Toxicity of α-Fe_2O_3 nanoparticles to *Artemia salina* cysts and three stages of larvae. Science of the Total Environment. 2017;**598**:847-855. DOI: 10.1016/j.scitotenv.2017.04.183

[18] Brundo MV, Pecoraro R, Marino F, Salvaggio A, Tibullo D, Saccone S, Bramanti V, Buccheri MA, Impellizzeri G, Scuderi V, Zimbone M, Privitera V. Toxicity evaluation of new engineered nanomaterials in zebrafish. Frontiers in Physiology. 2016;**130**. DOI: 10.3389/fphys.2016.0013

[19] Buccheri MA, D'Angelo D, Scalese S, Spano SF, Filice S, Fazio E, Compagnini G, Zimbone M, Brundo MV, Pecoraro R, Alba A, Sinatra F, Rappazzo G, Privitera V. Modification of graphene oxide by laser irradiation: A new route to enhance antibacterial activity. Nanotechnology. 2016;**27**:245704-245712. DOI: 10.1088/0957-4484/27/24/245704

[20] Pecoraro R, D'Angelo D, Filice S, Scalese S, Capparucci F, Marino F, Iaria C, Guerriero G, Tibullo D, Scalisi EM, Salvaggio A, Nicotera I, Brundo MV. Toxicity evaluation of graphene oxide and titania loaded nafion membranes in zebrafish. Frontiers in Physiology. 2017. DOI: 10.3389/fphys.2017.01039

[21] Pecoraro R, Marino F, Salvaggio A, Capparucci F, Di Caro G, Iaria C, Salvo A, Rotondo A, Tibullo D, Guerriero G, Scalisi EM, Zimbone M, Impellizzeri G, Brundo MV. Evaluation of chronic nanosilver toxicity to adult zebrafish. Frontiers in Physiology. 2017. DOI: 10.3389/fphys.2017.01011

[22] Gajdosíková A, Gajdosík A, Koneracká M, Závisová V, Stvrtina S, Krchnárová V, Kopcanský P, Tomasovicová N, Stolc S, Timko M. Acute toxicity of magnetic nanoparticles in mice. Neuro Endocrinology Letters. 2006;27(2):96-99

[23] Gad SC. Animal Models in Toxicology. London: CRC Press; 2014. p. 983

[24] Beasley A, Elrod-Erickson M, Otter RR. Consistency of morphological endpoints used to assess developmental timing in zebrafish (*Danio rerio*) across a temperature gradient. Reproductive Toxicology. 2012;34:561-567

[25] Kimmel CB, Ballard WW, Kimmel SR, Ullmann B, Schilling TF. Stages of embryonic development of the zebrafish. Developmental Dynamics. 1995;203:253-310

[26] Hill AJ, Teraoka H, Heideman W, Peterson RE. Zebrafish as a model vertebrate for investigating chemical toxicity: Review. Toxicological Science. 2005;86(1):6-19

[27] Ballard WW. Morphogenetic movements and fate maps of vertebrates. American Zoologist. 1981;21:391-399

[28] Belyaeva NF, Kashirtseva VN, Medvedeva NV, Khudoklinova YY, Ipatova OM, Archakov AI. Zebrafish as a model system for biomedical studies: Review. Biochemistry (Moscow) Supplement Series B Biomedical Chemistry. 2009;3(4):343-350

[29] Chen TH, Lin CY, Tseng MC. Behavioral effects of titanium dioxide nanoparticles on larval zebrafish (*Danio rerio*). Marine Pollution Bulletin. 2011;63:303-308

[30] Wood CM, Playle RC, Hogstrand C. Physiology and modeling of mechanisms of silver uptake and toxicity in fish. Environmental Toxicology and Chemistry. 1999;18:71-83

[31] Wang J, Zhu X, Zhang X, Zhao Z, Liu H, George R, Wilson-Rawls J, Chang Y, Chen Y. Disruption of zebrafish (*Danio rerio*) reproduction upon chronic exposure to TiO_2 nanoparticles. Chemosphere. 2011;83:461-467

[32] Paterson G, Ataria JM, Hoque ME, Burns DC, Metcalfe CD. The toxicity of titanium dioxide nanopowder to early life stages of the Japanese medaka (*Oryzias latipes*). Chemosphere. 2011;82:1002-1009

[33] Xu Z, Zhang YL, Song C, Wu LL, Gao HW. Interactions of hydroxyapatite with proteins and its toxicological effect to zebrafish embryos development. PLoS One. 2012;7(4):e32818

[34] Asharani PV, Lianwu Y, Gong Z, Valiyaveettil S. Comparison of the toxicity of silver, gold and platinum nanoparticles in developing zebrafish embryos. Nanotoxicology. 2010;5(1):43-54

[35] Paatero I, Casals E, Niemi R, Özliseli E, Rosenholm JM, Sahlgren C. Analyses in zebrafish embryos reveal that nanotoxicity profiles are dependent on surface-functionalization controlled penetrance of biological membranes. Scientific Reports. 2017 Aug 21;7(1):8423. DOI: 10.1038/s41598-017-09312-z

[36] Samaee SM, Rabbani S, Jovanovic B, Mohajeri-Tehrani MR, Haghpanah V. Efficacy of the hatching event in assessing the embryo toxicity of the nano-sized TiO_2 particles in zebrafish: A comparison between two different classes of hatching-derived variables. Ecotoxicology and Environmental Safety. 2015;116:121-128

[37] Ong KJ, Zhao X, Thistle ME, Mac Cormack TJ, Clark RJ, Ma G, Martinez-Rubi Y, Simard B, Loo JSC, Veinot JCG, Goss GG. Mechanistic insights into the effect of nanoparticles on zebrafish hatch. Nanotoxicology. 2014;8:295-304. DOI: 10.3109/17435390.2013.778345

[38] Embry MR, Belanger SE, Braunbeck TA, Galay-Burgos M, Halder M, Hinton DE, Léonard MA, Lillicrap A, Norberg-King T, Whale G. The fish embryo toxicity test as an animal alternative method in hazard and risk assessment and scientific research. Aquatic Toxicology. 2010;97(2):79-87

[39] OECD. Guideline for the Testing of Chemicals. Paris, France: Fish Embryo Toxicity (FET); 2013

[40] George S, Xia T, Rallo R, Zhao Y, Ji Z, Lin S, Wang X, Zhang H, France B, Schoenfeld D, Damoiseaux R, Liu R, Lin S, Bradley KA, Cohen Y, Nel AE. Use of a high-throughput screening approach coupled with in vivo zebrafish embryo screening to develop hazard ranking for engineered nanomaterials. ACS Nano. 2011;5(3):1805-1817

[41] Xu J, Zhang Q, Li X, Zhan S, Wang L, Chen D. The effects of copper oxide nanoparticles on dorsoventral patterning, convergent extension, and neural and cardiac development of zebrafish. Aquatic Toxicology. 2017;188:130-137

[42] Usenko CY, Harper SL, Tanguay RL. In vivo evaluation of carbon fullerene toxicity using embryonic zebrafish. Carbon, NY. 2007;45:1891-1898

[43] Zhu X, Tian S, Cai Z. Toxicity assessment of iron oxide nanoparticles in zebrafish (*Danio rerio*) early life stages. PLoS One. 2012;7(9):e46286. DOI: 10.1371/journal.pone.0046286

[44] Asharani PV, Serina NG, Nurmawati MH, Wu YL, Gong Z, Valiyaveettil S. Impact of multi-walled carbon nanotubes on aquatic species. Journal of Nanoscience and Nanotechnology. 2008;8:3603-3609. DOI: 10.1166/jnn.2008.432

[45] Fratoddi I, Venditti I, Cametti C, Russo MV. The puzzle of toxicity of gold nanoparticles. The case-study of HeLa cells. Toxicology Research. 2015;4:796-800. DOI: 10.1039/C4TX00168K

[46] Dobrovolskaia MA, McNeil SE. Immunological properties of engineered nanomaterials. Nature Nanotechnology. 2007;2:469-478. DOI: 10.1038/nnano.2007.223

[47] Patra HK, Banerjee S, Chaudhuri U, Lahiri P, Dasgupta AK. Cell selective response to gold nanoparticles. Nanomedicine. 2007;3:111-119. DOI: 10.1016/j.nano.2007.03.005

[48] Cho WS, Cho M, Jeong J, Choi M, Cho HY, Han BS, Kim SH, Kim HO, Lim YT, Chung BH, Jeong J. Acute toxicity and pharmacokinetics of 13 nm-sized PEG-coated gold nanoparticles. Toxicology and Applied Pharmacology. 2009;236:16-24. DOI: 10.1016/j. taap.2008.12.023

[49] Peng G, Tisch U, Adams O, Hakim M, Shehada N, Broza YY, Billan S, Abdah-Bortnyak R, Kuten A, Haick H. Diagnosing lung cancer in exhaled breath using gold nanoparticles. Nature Nanotechnology. 2009;4:669-673. DOI: 10.1038/nnano.2009.235

[50] Tedesco S, Doyle H, Blasco J, Redmond G, Sheehan D. Oxidative stress and toxicity of gold nanoparticles in *Mytilus edulis*. Aquatic Toxicology. 2010;100:178-186

[51] Pan Y, Neuss S, Leifert A, Fischler M, Wen F, Simon U, Schmid G, Brandau W, Jahnen-Dechent W. Size-dependent cytotoxicity of gold nanoparticles. Small. 2007;3:1941-1949. DOI: 10.1002/smll.200700378

[52] Zhang XD, Guo ML, Wu HY, Sun YM, Ding YQ, Feng X, Zhang LA. Irradiation stability and cytotoxicity of gold nanoparticles for radiotherapy. International Journal of Nanomedicine. 2009;4:165-173. DOI: 10.2147/IJN.S6723

[53] Sung JH, Ji JH, Park JD, Song MY, Song KS, Ryu HR, Yoon JU, Jeon KS, Jeong J, Han BS, Chung YH, Chang HK, Lee JH, Kim DW, Kelman BJ, Yu IJ. Subchronic inhalation toxicity of gold nanoparticles. Particle and Fibre Toxicology. 2011;14:16. DOI: 10.1186/1743-8977-8-16

[54] ACGIH. Threshold limit values and biological exposure indices for 1992-1993. In: American Conference of Governmental Industrial Hygienists. Cincinnati, OH; 1992

[55] Participants IRSIW. The relevance of the rat lung response to particle overload for human risk assessment: A workshop consensus report. Inhalation Toxicology. 2000;12: 1-17. DOI: 10.1080/08958370050029725

[56] Lee KP, Trochimowicz HJ, Reinhardt CF. Pulmonary response of rats exposed to titanium dioxide (TiO$_2$) by inhalation for two years. Toxicology and Applied Pharmacology. 1985;79:179-192. DOI: 10.1016/0041-008X(85)90339-4

[57] IARC. Titanium dioxide (IARC Group 2B), in IARC Monograph. Vol. 93. Lyon, France: International Agency for Research on Cancer; 2010

[58] Fako VE, Furgeson DY. Zebrafish as a correlative and predictive model for assessing biomaterial nanotoxicity. Advanced Drug Delivery Reviews. 2009;61:478-486. DOI: 10.1016/j.addr.2009.03.008

[59] Bourdineaud JP, Rossignol R, Brèthes D. Zebrafish: A model animal for analyzing the impact of environmental pollutants on muscle and brain mitochondrial bioenergetics. The International Journal of Biochemistry & Cell Biology. 2013;**45**:16-22. DOI: 10.1016/j.biocel.2012.07.021

[60] Salvaggio A, Marino F, Albano M, Pecoraro R, Camiolo G, Tibullo D, Bramanti V, Lombardo BM, Saccone S, Mazzei V, Brundo MV. Toxic effects of zinc chloride on the bone development in *Danio rerio* (Hamilton, 1822). Frontiers in Physiology. 2016;**7**(153). ISSN: 1664-042X. DOI: 10.3389/fphys.2016.00153

[61] Pecoraro R, Salvaggio A, Marino F, Caro GD, Capparucci F, Lombardo BM, Messina G, Scalisi EM, Tummino M, Loreto F, D'Amante G, Avola R, Tibullo D, Brundo MV. Metallic nano-composite toxicity evaluation by zebrafish embryo toxicity test with identification of specific exposure biomarkers. Current Protocols in Toxicology. 2017;**74**:1.14.1-1.14.13. DOI: 10.1002/cptx.34

4

Dose Duration Effects of 17-α Ethynylestradiol in Zebrafish Toxicology

Decatur Foster and Kim Hanford Brown

Abstract

Exposure of zebrafish to the synthetic estrogen 17-α ethynylestradiol (EE2) has been shown to cause a number of detrimental effects, including but not limited to feminization of male fish, reduced reproductive capabilities, and impaired embryonic development. This paper systematically reviews the effects of five environmentally relevant concentrations of EE2 on 12 measurements that are commonly selected when studying the effects of EE2 on zebrafish. Concentrations of 0.1 ng EE2/L, 1 ng EE2/L, 3 ng EE2/L, 10 ng EE2/L, 25 ng EE2/L, and 100 ng EE2/L were reviewed for their effects on sex ratio, vitellogenin induction, gonad morphology, spawning success, survival, bodily malformation, length/weight, swim-up success, fecundity, fertilization success, hatching success, and the reversibility of aforementioned effects. A greater occurrence of effects was observed as the dose of EE2 was increased, starting at exposure levels of 1 ng EE2/L. For exposures of 3 and 10 ng EE2/L, negative effects on sex ratio, morphology, and reproductive capabilities were reversible after zebrafish were able to recover in clean water for a period of time. Data for zebrafish exposed to 100 ng EE2/L was limited, as this concentration severely decreased survival.

Keywords: *Danio rerio*, zebrafish, 17-α ethynylestradiol, estrogen, endocrine disrupting chemical (EDC), toxicology

1. Introduction

Endocrine disrupting compounds (EDCs) are a class of chemical that have the ability to interfere with normal functions of the endocrine systems of living organisms. EDCs can affect organismal systems by mimicking, counteracting, or disrupting the synthesis and metabolism of endogenous hormones, as well as disturbing the synthesis of specific hormone receptors [1]. Among EDCs, estrogenic chemicals (ECs) are among the most extensively studied, primarily due to high levels of environmental contamination and a wide range of effects on aquatic ecosystems that have come to light over the past several decades. ECs can be found in many common household items, but their primary modes of entry into the environment are through wastewater effluent from municipal treatment plants, hospital effluent, and livestock activities [2].

Among ECs, 17 α-ethynylestradiol (EE2) is of particular concern, as it has been shown to be 10–50 times more potent in fish than naturally produced estrogen, due to its longer half-life and tendency to bio-concentrate in tissues [3]. EE2 is a derivative of the natural hormone estradiol (E2) and is commonly used as the bioactive estrogen for human oral contraceptive pills. In terms of frequency of use, oral contraceptives containing EE2 rank among the top 15 U.S. active pharmaceutical ingredients [4]. In addition to being utilized in human birth control, EE2 is also widely used in livestock to prevent pregnancy. Beyond contraceptive use, EE2 is utilized as a medicine for alleviating menopausal and postmenopausal syndrome symptoms, physiological replacement therapy for estrogen deficient states, treatment of prostatic cancer and breast cancer, and osteoporosis [1].

Human urine is considered the main source of EE2 contamination in the environment, as excess EE2 in the body is excreted in urine and enters aquatic systems through wastewater effluent release. Prior to EE2's excretion in urine, it is metabolized to become a biologically inactive, water-soluble sulfate or glucuronide conjugate [5]. Following excretion and subsequent transfer to wastewater treatment plants, EE2 may be activated into its free form due to bacterial modification. The activated EE2 remains relatively stable during the activated sludge process in sewage treatment plants, thus avoiding breakdown and elimination [6]. Because of EE2's highly stable molecular structure (**Figure 1**), it has become a widespread problem in the environment. Given its high resistance to degradation, and its tendency to be absorbed by organic matter, accumulate in sediment, and concentrate in biota, EE2 can cause significant issues for aquatic organisms and populations once present in the environment [7].

With a global human population of over 7 billion, it is estimated that approximately 700 kg/year of synthetic estrogens are released into the environment from contraceptive usage alone [8]. Environmental EE2 concentrations in water are highly variable, ranging from non-detectable levels to a maximum reported concentration of 830 ng/L in U.S. rivers [9]. As an example, a study in Washington State analyzed 266 surface water samples from lakes and streams in the Seattle area and detected EE2 in 66 samples, with a maximum concentration measuring 4 ng/L [10]. Other studies have observed concentrations of 42 ng/L in Canadian

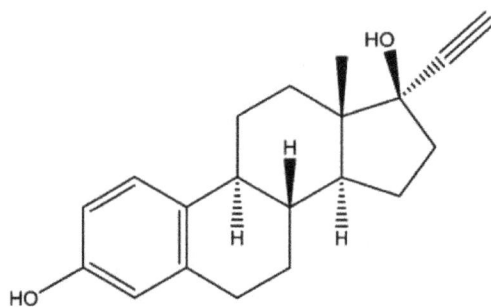

Figure 1. Chemical structure of 17-α ethynylestradiol (EE2).

sewage treatment effluent [11], while studies in Europe have found concentrations generally below 5 ng/L [12]. This has raised concern, as concentrations as low as 1 ng/L have been observed to affect offspring survival of adult male fish exposed to EE2 [13].

In some fish species, the binding affinity of EE2 to the estrogen receptor has been shown to be up to five times higher than E2 [3]. This higher receptor affinity indicates that EE2 can be a more potent estrogenic compound in terms of eliciting an estrogenic response, compared to naturally produced E2 [1]. Under environmental and laboratory conditions, EE2 has been reported to cause a wide variety of negative effects in multiple species of fish, including bias in the sex ratio toward females, decreased fertility and fecundity, vitellogenin induction in males, reduction of gonadal development, intersex, and impairment of reproductive behaviors [1, 7, 12–27].

Zebrafish (*Danio rerio*) are commonly used in laboratories to observe the effects of EE2 in aquatic models, as they exhibit most of the measurements that have been detected in a variety of fish species and have high gene ontology with humans [28]. Given their rapid development from fertilization to reproductive maturity in only three to four months, both short-term early life stage tests and chronic life-cycle tests can be conducted in a relatively short amount of time [19]. The short life cycle is also beneficial when studying developmental and reproductive effects of endocrine disrupting compounds [20]. Their ability to breed year round makes zebrafish ideal for studies observing fecundity and fertility. Furthermore, zebrafish produce a large number of transparent eggs per spawn, which is preferable when collecting both quantitative and morphological data. Finally, zebrafish are well studied; embryogenesis in this species has been researched in detail, and the entire zebrafish genome has been published, allowing for in-depth genetic comparison and analysis [23].

In this paper, we focus on reviewing the impact that EE2 has on 12 measurements of fitness that are commonly selected when studying the effects of EE2 on zebrafish. They include: (1) skewed sex ratios from male to female; (2) the induction of vitellogenin (VTG) in male fish (an egg yolk precursor protein normally expressed only in females); (3) gonad morphology (undeveloped gonads, mature ova/testes or intersex - see **Figure 2**); (4) spawning success (onset of spawning and number of successful spawns); (5) survival; (6) bodily malformation; (7) length/weight; (8) swim-up success (successful inflation of the swim bladder by day 7 post fertilization) (9) fecundity (number of eggs per spawn); (10) number of viable eggs per spawn (fertilization success); (11) number of hatched eggs per spawn; and (12) reversibility of effects (the ability of these 11 measurements to return to control levels after a period of depuration). This review will help summarize the vast amount of zebrafish research that has been published over the past two decades pertaining to EE2 exposure.

We chose five concentrations of EE2 that were most commonly used by researchers, all of which are environmentally relevant: 0.1 ng/L, 1 ng/L, 3 ng/L, 10 ng/L, 25 ng/L, and 100 ng/L. These studies observed exposure periods of 5–180 days, followed by a depuration period of 25–150 days in order to test for reversibility of effects (**Figure 3**). Studies that did not begin exposure at day 1 (i.e. partial lifecycle exposures) were excluded from this review. Furthermore, effects on second-generation exposure fish are not reported in this review. All findings reported in this review were deemed statistically significant by the original authors, as compared to control tests, unless otherwise noted.

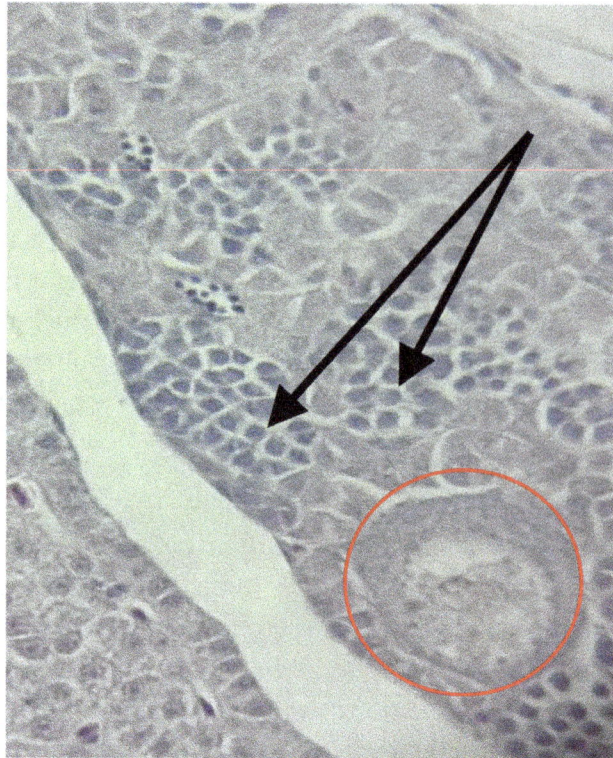

Figure 2. Histology image of intersex tissue from 8-week-old zebrafish larva (red circle indicating an oocyte and black arrows indicating testicular tissue) [17].

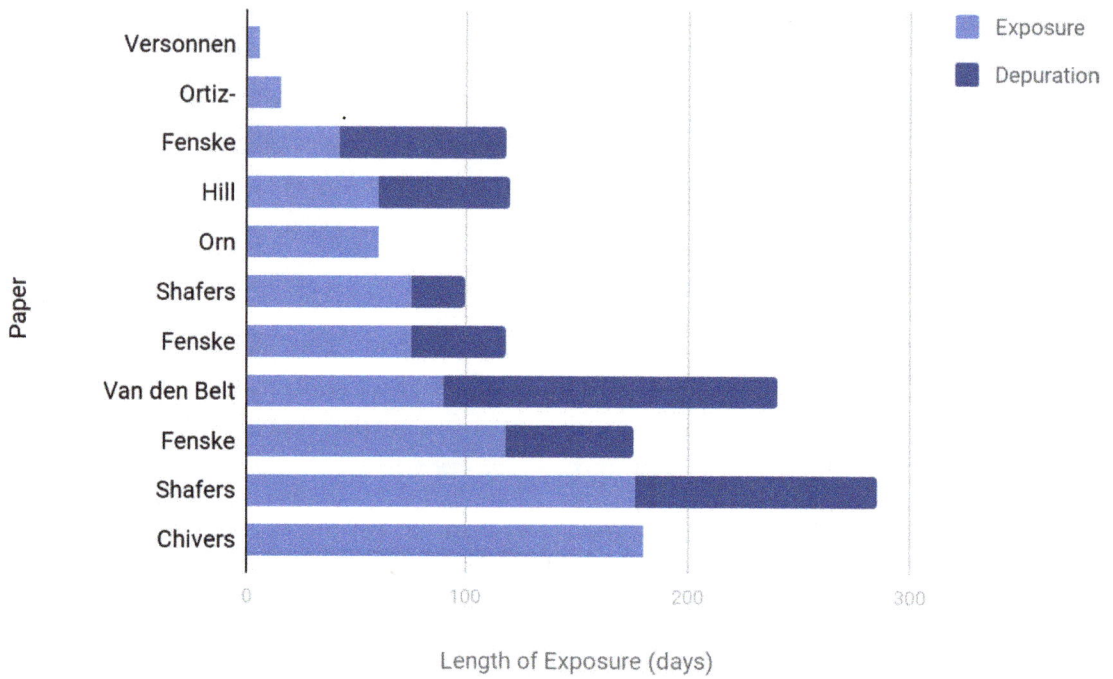

Figure 3. Length of EE2 exposure and depuration observed in each zebrafish study evaluated.

2. Control experiment parameters

It is important to consider that a number of factors outside of exposure to EE2 can affect the 12 measurements reviewed in this paper. The age and size of fish, interval between successful spawns, light cycle, and diet all have an effect on quality and quantity of egg production. There is often little continuity in these environmental factors between aquatic laboratories, which can lead to differences in control outcomes. One additional complication with cross evaluation of studies is the variety of solvents used between different laboratories to dissolve EE2 for exposure trials. Acetone, methanol, and ethanol are most commonly used in the reviewed papers we evaluated, but other solvents have been reported, which may have differing effects on organismal physiology.

Laboratory zebrafish typically attain sexual maturity in the 3rd month of their development, but initial spawns can be observed in fish at ages as young as 2.5 months. Once sexual maturity is reached, prime reproductive performance is maintained for several months, but decreases with advancing age. Optimal zebrafish reproduction through natural mating occurs when the fish are aged 6 months to 1 year [24].

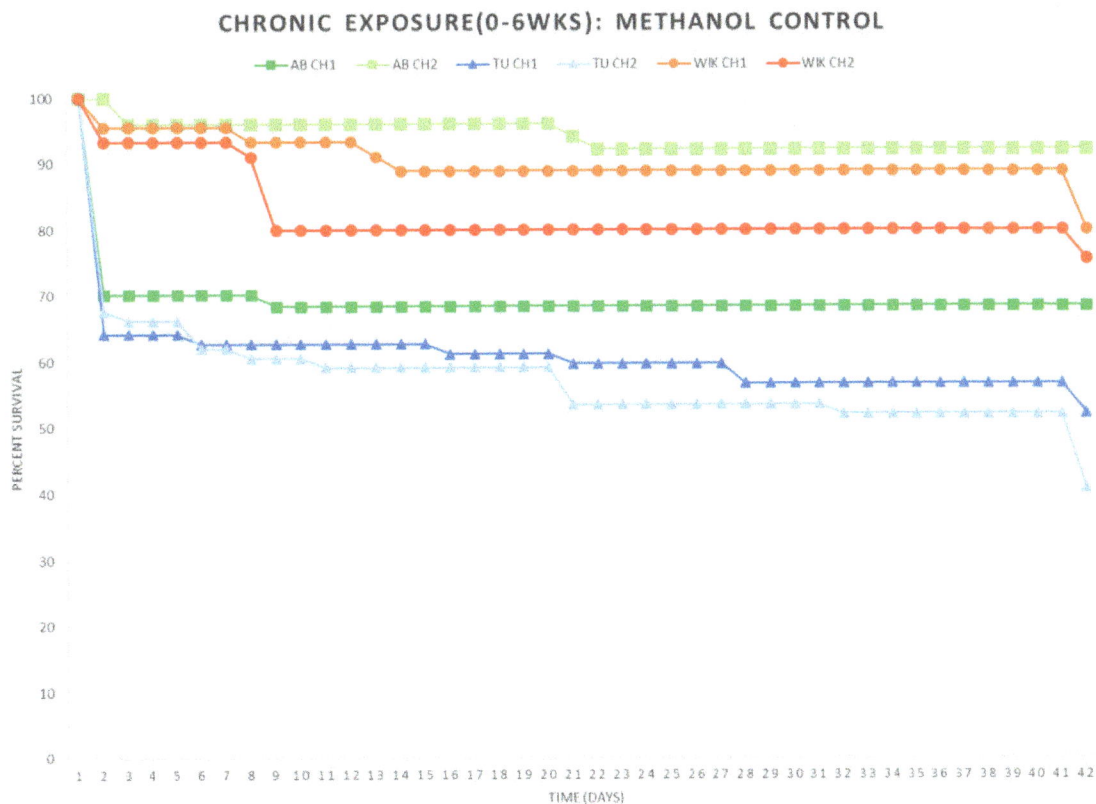

Figure 4. Survival from fertilization to 6 weeks of three different strains of zebrafish (AB, TU, WIK) under control conditions (no EE2 present). Survival of two cohorts per strain (CH1 and CH2) is shown, exhibiting substantial intra- and inter-strain variation [17].

Finally, and perhaps most importantly, few papers specify which strain of zebrafish are used. Zebrafish researchers typically report that "wild type" zebrafish are used, but fail to specify which "wild type" they are referring to. Many different laboratory "wild type" strains are available, including AB, Tuebingen (TU), WIK, and Tupfel long fin (TL), among others. The specific "wild type" strain used could have an impact on control and exposure outcome, given the extensive genetic diversity between laboratory strains [25]. Among the array of laboratory strains available, the three most common laboratory strains include AB, TU, and WIK [17]. Each of these strains differs in their initial method of establishment, historic degree of selective breeding, and genetic bottlenecks that likely affect physiological performance under exposure. **Figure 4** represents such strain variability in survival of zebrafish, based solely on vehicle control conditions.

3. Exposure to 0.1 ng EE2/L

Overall, exposure to 0.1 ng EE2/L appears to have little or no effect on zebrafish. Two studies evaluated concentrations of EE2 at this level with no detrimental effects observed. Zebrafish in these studies were evaluated for a period of 90 days [18] and a period of 177 days [19].

3.1. Sex ratio

After 90 days of exposure, 40% of zebrafish were female, while 40% had undeveloped gonads [18]. This did not significantly differ from control ratios.

3.2. VTG levels

After 90 days of exposure, VTG was not detected in male fish [18].

3.3. Bodily malformation

After 90 days of exposure, no bodily malformation was observed in zebrafish [18].

3.4. Length/weight

After 90 days of exposure, the total body length and weight of zebrafish was not significantly different than that of the control fish [18].

3.5. Fecundity

After 177 days of exposure (at 0.05 ng/L), there was no statistically significant difference in number of eggs produced per day between exposure and control zebrafish. Exposure fish produced 32.6 eggs per day [19].

3.6. Viable eggs

After 177 days of exposure (at 0.05 ng/L), there was no statistically significant difference in the number of successfully fertilized eggs between exposure and control zebrafish. Exposure fish had a fertilization success rate of 91.6% [19].

4. Exposure to 1 ng EE2/L

Exposure to 1 ng EE2/L impacted onset of spawning, fecundity, as well as the number of viable eggs per spawn. Spawning did not occur during the exposure period for fish that were exposed for 177 days. The percentage of eggs laid per day by exposed fish was reduced by 24%, while the percentage of viable eggs reduced from approximately 95% in control fish to 41–51% in exposure fish. Three papers were reviewed at this concentration of EE2, observing effects for a period of 60 days [20], 90 [18], and 177 days [19].

4.1. Sex ratio

After 60 days of exposure, 77% of zebrafish were female, while 5% had undeveloped gonads [20]. After 90 days of exposure, 50% of zebrafish were female, while 40% had undeveloped gonads [18]. None of these data points were deemed statistically significant.

4.2. VTG induction

After 90 days of exposure, VTG in males was measured at approximately 150 ng/mg, which was not deemed statistically significant compared to the control [18].

4.3. Spawning success

After 75 and 177 days of exposure, there was a delay in onset of spawning [19]. For the group that was exposed for 177 days, spawning occurred at day 120, while control fish spawned at day 112 [19]. These data points were not deemed statistically significant.

4.4. Survival

After 6 weeks of exposure, there was no significant difference in survival between exposure fish and control fish. The greatest loss occurred within the first 10 days post fertilization, which was attributed to normal larval mortality. **Figure 5** shows this survival curve, with other major losses shown at the end of the 6 weeks attributed to miscounts and/or cannibalism [17]. There was no statistically significant difference in survival between exposure fish and control fish at day 42 [19], day 60 (survival rate of 60% was reported) [20], or day 75 of exposure [19].

4.5. Bodily malformation

After 90 days of exposure [18] and 180 days of exposure [17], there was no bodily malformation.

4.6. Length/weight

After 60 days of exposure [20], and 90 days of exposure [18], the total body length and weight of exposure fish was not significantly different than that of the control fish.

4.7. Swim up success

There was no statistically significant difference in swim up success between exposure and control fish [26].

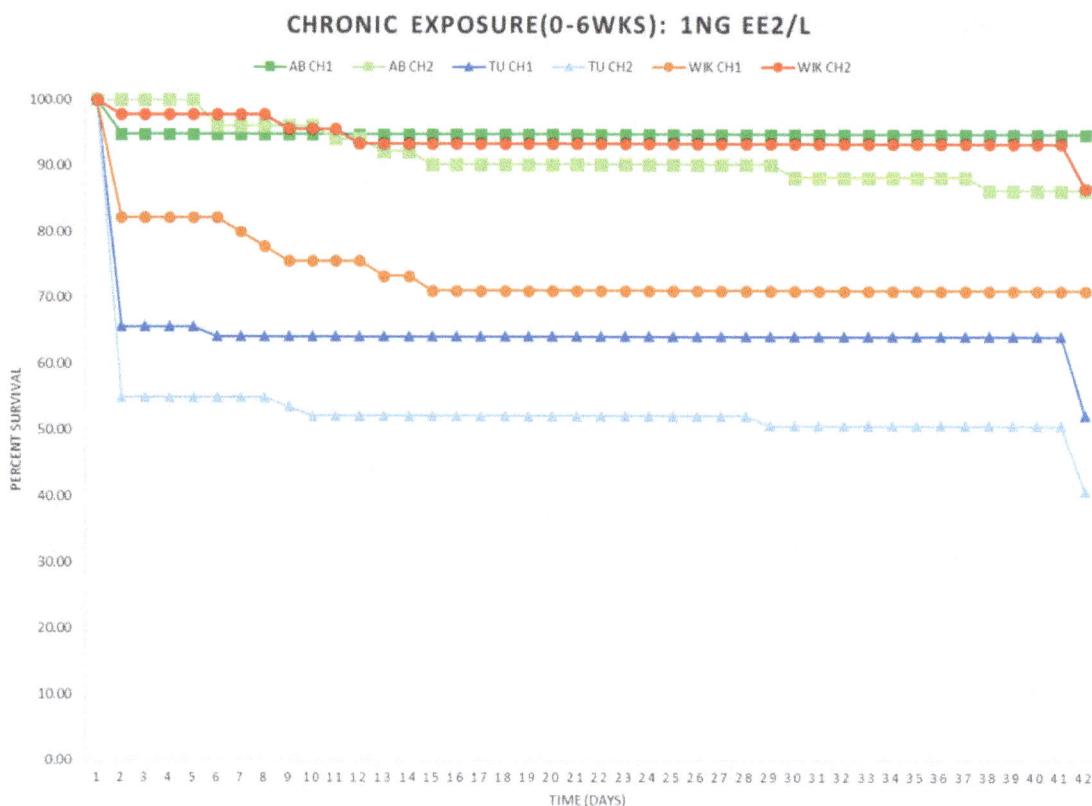

Figure 5. Percent survival from 0 days post fertilization to 6 weeks for 1 ng EE2/L treatment groups of three zebrafish strains: AB (green), TU (blue), and WIK (orange). Two cohorts per strain (CH1 and CH2) are shown. Results were not statistically different from control cohorts [17].

4.8. Hatching success

There was no statistically significant difference in hatching success between exposure and control fish [26].

4.9. Fecundity

When fish had been exposed for 75 days, fecundity was not affected [19]. After 177 days of exposure, the number of eggs laid per day by exposure fish was reduced to 23.3 eggs per day, as compared to 30.4 eggs per day in control fish [19]. This was deemed a statistically significant difference.

4.10. Viable eggs

When fish had been exposed for 75 days, the percentage of viable eggs was reduced from approximately 95% in control fish to 41% in exposed fish [19]. After 177 days of exposure, the percentage of viable eggs was reduced from 95% in control fish to 51.8% in exposed fish [19]. These data points were deemed statistically significant.

5. Exposure to 3 ng EE2/L

Exposure to 3 ng EE2/L significantly impacted VTG levels and spawning success in zebrafish. VTG levels were greater in exposed fish than in control fish after 75 days of exposure. Spawning did not occur during the exposure period for fish that were exposed for 188 days. One paper was reviewed at this concentration of EE2, observing effects for a period of 42 days, 75 days, and 118 days [21].

5.1. Sex ratio

After 42 days of exposure, the sex ratio of exposed fish was unaffected [21].

5.2. VTG levels

After 42 days of exposure, no difference was observed in the body homogenate VTG concentrations between the EE2 exposed and control fish. A range of 0.05 to 7.75 µg/ml was detected [21]. After 75 days of exposure, mean plasma VTG concentration in exposed fish was elevated over control values to a level that was deemed statistically significant. Inter-individual variation was high, with VTG concentration in exposed fish ranging between 14.76 and 1356.21 µg/ml. This variation could have been caused by the selection sample, which was of unknown sex [21], as females are known to show less increase in VTG compared to males. After 118 days of exposure, male fish had significantly increased levels of VTG [21].

5.3. Gonad morphology

After 42 days of exposure, the histological appearance of the ovaries was not different from the control fish. However, testes were less developed than in control fish; only two out of nine male fish possessed mature testes, while the other seven fish had immature testes [21]. After 75 days of exposure the ovarian histology did not differ from the control group [21]. After 118 days of exposure, all 27 individuals examined possessed ovaries, and none of the fish had gonads of testicular morphology. Both mature and immature ovaries were present: thirteen individuals had developed ovaries, with all oocyte stages and post-ovulatory follicles, while 14 fish had immature ovaries with exclusively oogonia and primary growth stage oocytes. In the 13 fish with mature ovaries, oocyte maturation was less progressed than in mature ovaries of control fish [21].

5.4. Spawning success

When fish were exposed for 42 days, the initiation of spawning was not altered. The first spawning event occurred at 83 days post fertilization (DPF), while control fish started spawning between 80 and 82 DPF [21]. However, when fish were exposed for 118 days, they did not spawn during the exposure period [21].

5.5. Survival

After 28 days of exposure, survival of exposed fish was 56–84%, which was not statistically different than control fish [21].

5.6. Length/weight

After 28 days of exposure, exposed fish were longer in length compared to control fish [21]. However, after 45 days of exposure and 75 days of exposure, there was no statistically significant difference [21].

5.7. Hatching success

The number of successfully hatched fish per spawn was not affected [21]

5.8. Fecundity

When fish were exposed for 42 days, fecundity was not statistically different compared to control fish [21].

5.9. Viable eggs

When fish were exposed for 42 days, there was no statistically significant difference in number of viable eggs between exposed and control fish (85.3% fertilization compared to 90.1% in the control) [21].

6. Exposure to 10 ng EE2/L

Exposure to 10 ng EE2/L significantly impacted sex ratio (up to 100% female), VTG levels (increased), gonad morphology (no mature ovaries), spawning success (delay in onset of spawning), length/weight of zebrafish (reduced), fecundity (reduced) and number of viable eggs (reduced). Six papers were reviewed at this concentration of EE2, observing effects for a period of 7 and 14 days [23], 60 days [19, 21], 90 days [18], 177 days [19], and 180 days [17].

6.1. Sex ratio

After 60 days of exposure, 18% of zebrafish were female, while 82% had undeveloped gonads, which was found to be statistically significant as compared to control fish [20]. A different study reported that 100% of zebrafish were female with well-defined ovaries after 60 days of exposure (compared to the mean percentages of male and female zebrafish of 33% and 67% in control) [22]. After 90 days of exposure, a sex ratio of 30% female and 70% undeveloped gonads was reported [18]. After 180 days exposure, 80% of zebrafish were female, which was statistically higher than the control ratio [17].

6.2. VTG levels

After 7 days of exposure, plasma VTG levels were measured at 760 ug/mg protein, while at day 14, levels were measured at 1272 ug/mg protein [23]. Two other papers reported that VTG induction was observed after 60 days of exposure [20], with levels being higher in exposure fish (4900 μg VTG/g) compared to control (0.5 μg/g) [22]. After 90 days of exposure, one study found VTG levels of approximately 575 μg/g [18]. All of these data points were deemed statistically significant.

6.3. Gonad morphology

After 60 days of exposure, 16 out of 20 zebrafish possessed undeveloped gonads, as compared to only one fish in the control group with undeveloped gonads. In these fish, only a small mass of primordial gonadal cells were located at the genital ridge lining the edges of the liver and swim bladder. Thus, although gonadal tissue was present, gonads were classified as undeveloped when no discernable cells characteristic of either sex were observed [20]. After 177 days of exposure, all individuals displayed gonads with ovarian morphology, but no mature ovaries. Three fish possessed ovaries containing vitellogenic and mature oocytes, while the ovaries of the remaining 24 fish contained immature pre-vitellogenic oocytes only, mostly at the perinucleolar stage and in a few cases at the cortical alveolar stage. Fish with testes were not found among all 27 individuals [19].

6.4. Spawning success

When fish were exposed for 75 days, spawning was delayed [19]. When fish were exposed for 90 days, there was a reduction in the number of spawning females within 3 separate spawning periods that were observed [18]. No mating behavior or spawning occurred during a 177 day exposure [19].

6.5. Survival

After 6 weeks of exposure, there was no significant difference in survival between exposure fish and control fish. The greatest loss occurred within the first 10 days post fertilization, which was attributed to normal larval mortality. **Figure 6** shows this survival curve, with other major losses shown at the end of the 6 weeks attributed to miscounts and/or cannibalism [17]. After 60 days of exposure, a 42% survival rate was observed, which was not statistically different than control fish [20]. After 75 days of exposure, survival of exposure fish was slightly lower than in control fish, but not statistically significant [19].

6.6. Bodily malformation

After 90 days [18] and 180 days [17] of exposure, no bodily malformation was observed.

6.7. Length/weight

After 42 days of exposure, no difference in length/weight was observed [19]. After 60 days of exposure, body length of females had decreased compared to control [20]. Separate papers

CHRONIC EXPOSURE(0-6WKS): 10NG EE2/L

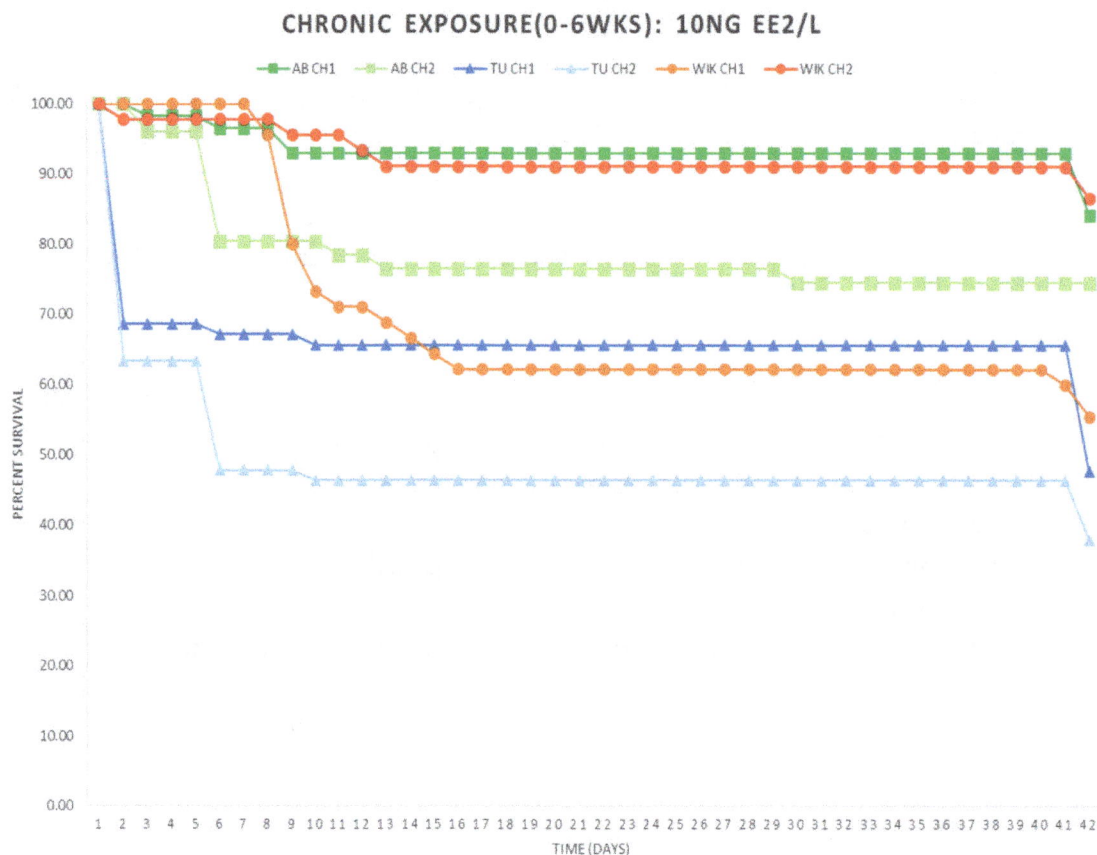

Figure 6. Percent survival from 0 days post fertilization to 6 weeks for 10 ng EE2/L treatment groups of three zebrafish strains: AB (green), TU (blue), and WIK (orange). Two cohorts per strain (CH1 and CH2) are shown. Results were not statistically different from control cohorts [17].

observing 75 and 90 day exposure periods both found that body length of exposed fish was reduced compared to control fish [17–18]. These data points were deemed statistically significant.

6.8. Swim-up success

No significant difference was observed [26].

6.9. Hatching success

No significant difference was observed [26].

6.10. Fecundity

After 75 days of exposure, fecundity was not affected [19]. However, after 90 days of exposure, total egg production was significantly reduced, down from approximately 70 eggs per female in the control group to approximately 45 eggs per female in the exposed group [18].

6.11. Viable eggs

After 75 days of exposure, the number of viable eggs was reduced from approximately 95% in the control group to 41% in the exposed group [19]. This was deemed statistically significant.

7. Exposure to 25 ng EE2/L

Exposure to 25 ng EE2/L impacted sex ratio (increase in % females) VTG levels (increased), gonad morphology (no developed gonads), spawning success (absence of activity), and length/weight of zebrafish (reduced). Two papers were reviewed at this concentration of EE2, observing effects for a period of 90 days [18] and 180 days [17].

7.1. Sex ratio

After 180 days exposure, approximately 75% of zebrafish were female, which was statistically higher than the control ratio [17].

7.2. VTG levels

After 90 days of exposure, VTG levels in males were observed at approximately 1100 ng/mg, which was significantly higher than VTG levels in control males [18].

7.3. Gonad morphology

After 90 days of exposure, 100% of zebrafish had undeveloped gonads [18].

7.4. Spawning success

After 90 days of exposure, there was a complete absence of the spawning activity [18].

7.5. Survival

After 6 weeks of exposure, there was no significant difference in survival between exposure fish and control fish. The greatest loss occurred within the first 10 days post fertilization, which was attributed to normal larval mortality. **Figure 7** shows this survival curve, with other major losses shown at the end of the 6 weeks attributed to miscounts and/or cannibalism [17]. Between 6 weeks and 6 months of exposure, exposed fish exhibited survival rates of 70–90% (**Figure 8**), which was not statistically different from control values. Phenotypically, fish had a similar appearance to other treatment groups, despite survival patterns not demonstrating a similar pattern.

7.6. Bodily malformation

After 90 days of exposure, 17% of zebrafish suffered from pericardial edema, and 51% exhibited lordosis and/or scoliosis [18]. After 180 days of exposure, edema in body cavity (**Figure 9**) and bulging eye was observed [18].

7.7. Length/weight

After 90 days of exposure, the total body length and weight of exposed fish was significantly lower than that of the control fish [18].

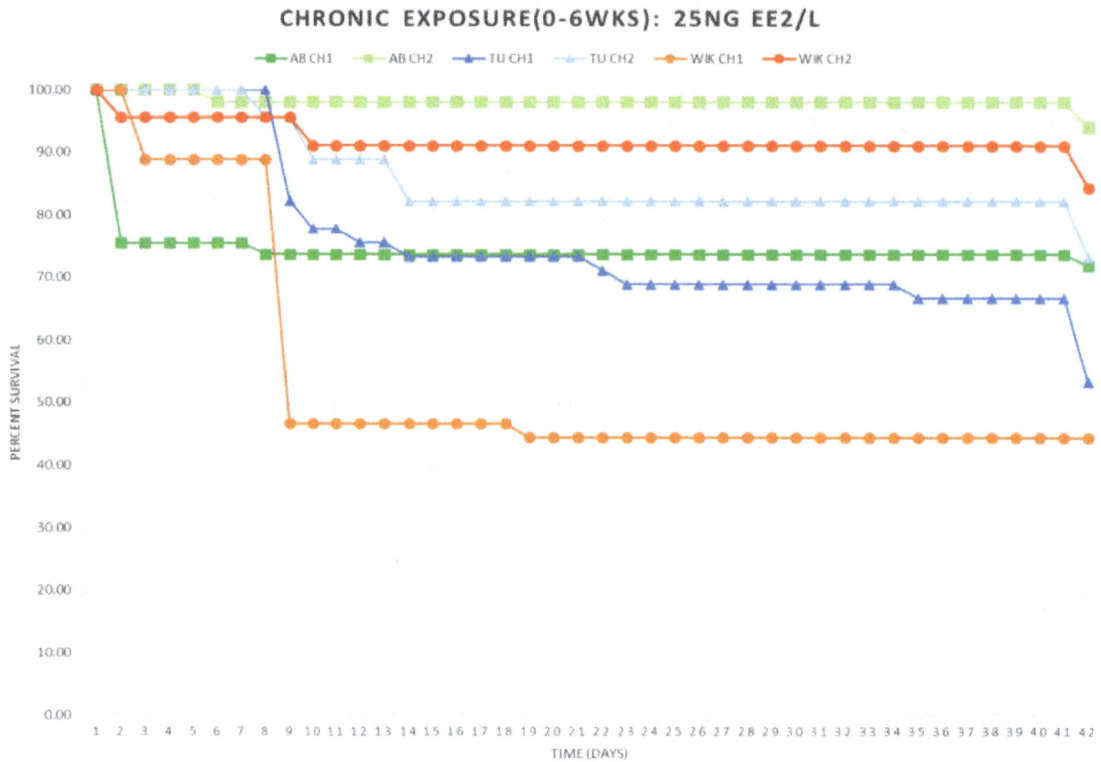

Figure 7. Percent survival from 0 days post fertilization to 6 weeks for 25 ng EE2/L treatment groups of three zebrafish strains: AB (green), TU (blue), and WIK (orange). Two cohorts per strain (CH1 and CH2) are shown. Results were not statistically different from control cohorts [17].

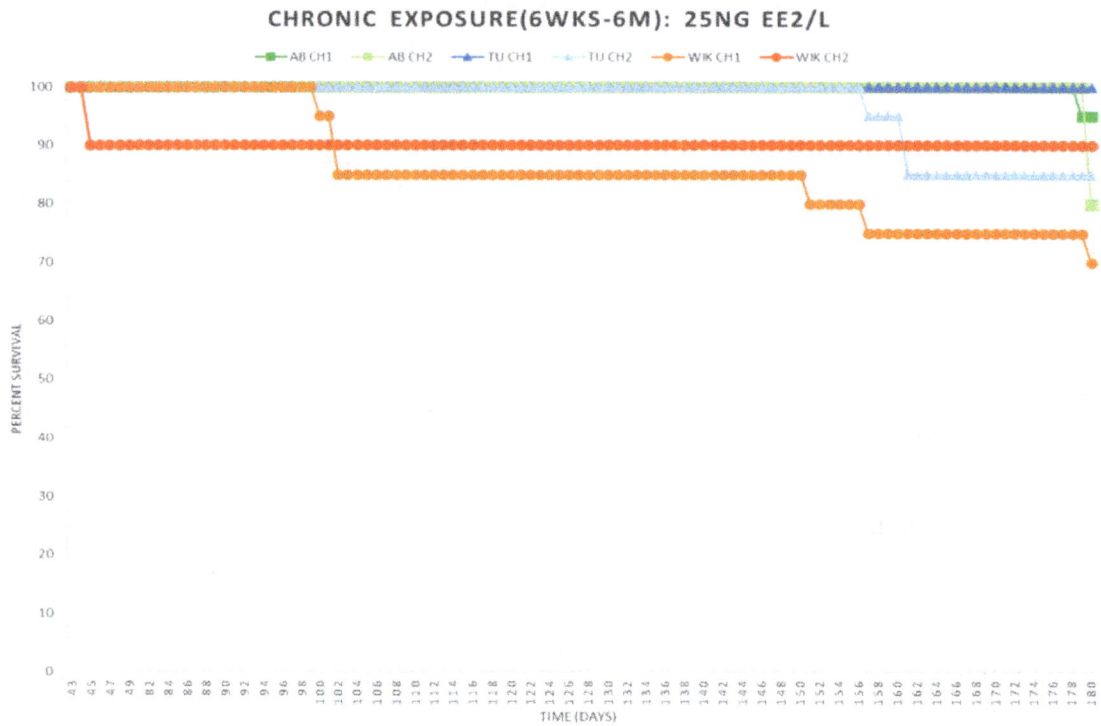

Figure 8. Percent survival from 6 weeks to 6 months for 25 ng EE2/L treatment groups of three zebrafish strains: AB (green), TU (blue), and WIK (orange). Two cohorts per strain (CH1 and CH2) are shown. Results were not statistically different from control cohorts [17].

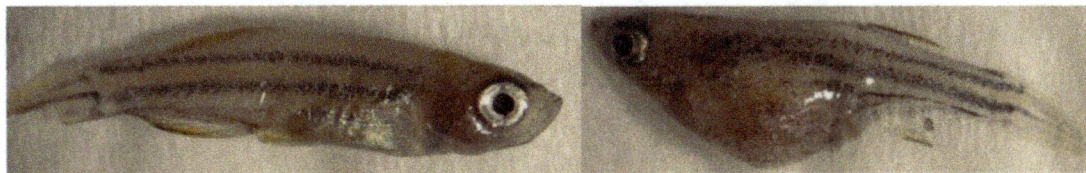

Figure 9. Image via stereoscopic microscope camera of control zebrafish larva at 8 weeks (left) and 25 ng EE2/L exposed larva exhibiting edema in body cavity (right) [17].

8. Exposure to 100 ng EE2/L

Exposure to 100 ng EE2/L impacted survival (decreased), swim-up success (decreased), and hatching success (delayed and decreased). Three papers were reviewed at this concentration of EE2, observing effects for a period of 120 hours [26], 14 days [22], and 60 days [20].

8.1. VTG levels

After 60 days of exposure, VTG induction in males was observed [20].

8.2. Survival

After 14 days of exposure, there was 0% survival [22]. Another paper observed that after 60 days of exposure, less than 10% of exposed fish survived [20].

8.3. Swim-up success

Swim up success was reduced to 60%, compared to 91% in control [26]. This was deemed statistically significant.

8.4. Hatching success

Hatching success was significantly reduced to 67%, compared to 95% in control. Hatching was also delayed compared to control (50% at 72 hours post fertilization (HPF) compared to 100% in control at 72HPF) [26].

9. Reversibility of Effects

When considering reversibility of effects, only measurements that authors had deemed significantly affected by exposure (see **Table 1**) are considered below. A statistically significant difference between exposed fish and control fish indicates that the aforementioned effect could not be reversed after a period of depuration. For reversibility to successfully occur, there must be no difference between control and exposure groups following depuration. This indicates that the measurement has reached control levels after a period of exposure followed by depuration. Results of this section are summarized in **Table 2**.

	0.1 ng EE2/L	1 ng EE2/L	3 ng EE2/L	10 ng EE2/L	25 ng EE2/L	100 ng EE2/L
Sex ratio (% female)	no difference	no difference	no difference	↑	↑	
VTG levels in males	no difference	no difference	↑	↑	↑	↑
Abnormal gonad morphology			X	X	X	
Spawning success		↓	↓	↓	↓	
Survival		no difference	no difference	no difference	no difference	↓
Bodily malformation	no difference	no difference		no difference	X	
Length/weight	no difference	no difference	no difference	↓	↓	
Swim-up success		no difference		no difference		↓
Hatching success		no difference	no difference	no difference		↓
Fecundity	no difference	↓	no difference	↓		
Viable eggs	no difference	↓	no difference	↓		

Blank spaces indicate that the measurement was not tested at a specific level. An 'X' indicates that the measurement was observed. A "↓" indicates that there was a significant decrease, while a "↑" indicates there was a significant increase in the measurement.

Table 1. Observed effects of acute exposure to EE2 in zebrafish that were deemed statistically significant at varying concentrations.

	3 ng EE2/L	10 ng EE2/L	25 ng EE2/L
Sex ratio (% female)	N/A	X	X
VTG levels in males	X	not reversible	
Abnormal gonad morphology	X	X	
Spawning success	X	X	
Fecundity	N/A	X	
Viable eggs	not reversible	not reversible	

Blank spaces indicate that the factor was not tested at a specific level. An 'N/A' means that the factor was not significantly affected by exposure, thus could not be measured for reversibility. An 'X' indicates that the measure was reversible.

Table 2. Reversibility of effects that were observed after acute exposure to EE2 in zebrafish, followed by a period of depuration.

9.1. Sex ratio

9.1.1. 10 ng/L

After 60 days of exposure and 60 days in clean water, 25% of zebrafish were female and 75% male [20]. After 90 days of exposure and 150 days in clean water, 65% of zebrafish were female and 35% male [18]. This was not significantly different than control values and indicates a reversal of effects.

9.1.2. 25 ng/L

After 90 days of exposure and 150 days in clean water, 35% of zebrafish were female and 65% male [18]. This was not significantly different than control values, and indicates a reversal of effects.

9.2. VTG levels

9.2.1. 3 ng/L

After 42 days of exposure and 76 days in clean water, there was no significant difference in VTG levels of males, which indicates a reversal of effects. After 118 days of exposure and 58 days in clean water, plasma VTG concentrations were approaching control levels in most exposed fish, and the agreement between the gonadal sex and the VTG level of individual fish was much higher than at 118 DPF [21].

9.2.2. 10 ng/L

After 177 days of exposure and 108 days in clean water, mean plasma vitellogenin levels were significantly higher at 6.7 ug VTG/ml plasma in male fish, compared with plasma VTG concentrations on average below detection limit in control males [19].

9.3. Gonad morphology.

9.3.1. 3 ng/L

After 42 days of exposure and 76 days in clean water, 17 out of 30 zebrafish possessed ovaries and 13 possessed testes. The histological appearance of the ovaries showed pronounced inter-individual variation: in 11 phenotypic females, mature ovaries were observed, whereas in six of the 17 ovary containing individuals, immature ovaries were found, containing oogonia and primary growth stage oocytes but no vitellogenic or mature oocyte. The occurrence of two types of ovaries was visible macroscopically during dissection of the fish: while in the case of mature ovaries numerous eggs were externally visible, immature ovaries appeared small, with no macroscopically recognizable substructure. The differentiated, mature testes contained numerous spermatozoa. One male had testis-ova, with a low number of primary growth stage oocytes dispersed in differentiated testicular tissue [21]. After 42 days of exposure and 134 days in clean water, 13 out of 29 fish of this treatment possessed normally differentiated testes, with all sperm stages being present. The amount of sperm cells among spermatocytes varied between individual males. The remaining 16 fish examined showed gonads with ovarian morphology, whereby eight individuals had mature and the other eight had immature ovaries [21]. After 118 days of exposure and 58 days in clean water, six out of 27 fish had fully differentiated testes, and one male displayed testis-ova. The other 20 fish possessed ovaries, of which 19 were developed as ovaries and one ovary was immature, containing oogonia and primary growth stage oocytes but with no further stages of oocyte maturation present [21].

9.3.2. 10 ng/L

After 60 days of exposure and 60 days in clean water, none of the exposed fish possessed undeveloped gonads or ovatestes/testis-ova, as was observed at 60 days post hatch (dph) [20]. After 177 days of exposure and 108 days in clean water, 5 out of 20 fish displayed gonads with the morphology of mature testes, and the remaining 15 fish had gonads with the morphology of mature ovaries [19].

9.4. Spawning success

9.4.1. 3 ng/L

After 118 days of exposure, spawning resumed after 22 days in clean water. This was a significant six week delay in the initiation of spawning compared to temporary, acute exposures performed only during the early life history stage [21]. However, the absence of spawning activity during exposure was successfully reversed.

9.4.2. 10 ng/L

After 177 days of exposure, spawning resumed after approximately 75 days in clean water [19]. The absence of spawning activity during exposure was successfully reversed.

9.5. Fecundity

9.5.1. 10 ng/L

After 75 days of exposure and 25 days in clean water, female fish laid 95 eggs per day [19]. After 177 days of exposure and 108 days in clean water, female fish laid 20.7 eggs per day, as compared to control fish that laid 30.4 eggs per day [19]. The authors did not indicate if there was a statistically significant difference between these two values.

9.6. Viable eggs

9.6.1. 3 ng/L

After 118 days of exposure and 58 days in clean water, fertilization success was significantly reduced from 91% in the control group to 21.7% in the exposed fish [21].

9.6.2. 10 ng/L

After 60 days of exposure and 60 days in clean water, 82% of eggs were viable, which was significantly reduced compared to 90% for control fish [20]. After 75 days of exposure and 25 days in clean water, the success rate was significantly reduced to 41% [19]. After 177 days of exposure and 252 days clean water, only 3% of eggs were successfully fertilized [19].

10. Conclusion

Table 1 summarizes the effects of the 6 concentrations of EE2 on each of the 12 measurements. As expected, we see a greater occurrence of effects as the dose of EE2 is increased, starting at exposure levels as low as 1 ng/L EE2. Data for 100 ng EE2/L exposure is limited, as zebrafish do not often survive at this concentration. It can be concluded that the impact of EE2 on zebrafish sexual development and reproductive functions, as well as the reversibility of effects, varies with exposure concentration, timing, and duration. In studies that had a short duration of EE2 exposure, negative effects persisted only if the exposure occurred during sexual differentiation and gametogenesis [20]. Male zebrafish pass through a stage of juvenile hermaphroditism, developing juvenile ovaries which are then transformed into testes between 20 and 60 dph [22]. This gonad transition stage is critical with respect to persisting effects of EE2 exposure on reproduction. Additionally, these experiments also illustrate the plasticity of gonadal differentiation and development in zebrafish, as a period of depuration was able to reverse many of the observed effects [20].

Though we can draw numerous conclusions from this data, a number of questions remain. First, what could be considered the toxic lethal dose (LD50) for zebrafish? It appears that the LD50 may be dependent on not only concentration, but length of exposure as well. Second, are there any differences in effects between strains? Deviations observed between evaluated studies could indicate the presence of strain specific effects. This is difficult, if not impossible, to determine since the preponderance of published papers do not identify which specific strain of zebrafish is being utilized. Without this information, reproducibility of the experiments is difficult as different strains may result in different outcomes. Additional factors, including zebrafish age and general husbandry techniques may differ between experiments and also compound the inability to replicate data.

Despite these assessments of life history exposure periods, there is a significant gap in our knowledge when it comes to chronic exposures longer than one life cycle. Studies on non-zebrafish species have shown that continuous exposure over extended generations can drive fish populations to near extinction [27]. While we have observed these long-term exposure effects, we still do not understand the mechanisms that underlie the dramatic population crashes. This knowledge gap could be filled using zebrafish, since they have a relatively short lifespan (less than three years) and generation time (~3 months). These characteristics make zebrafish an ideal model to test for the possibility of compounding effects over multiple generations. Such future studies will be particularly important for understanding long-term environmental impacts that result from continuous exposure of native populations and the mechanisms that cause such dramatic population declines.

Author details

Decatur Foster and Kim Hanford Brown*

*Address all correspondence to: kibr2@pdx.edu

Portland State University, Portland, OR, USA

References

[1] Aris AZ, Shamsuddin AS, Praveena SM. Occurrence of 17α-ethynylestradiol (EE2) in the environment and effect on exposed biota: a review. Environment International. 2014; **69**:104-119

[2] Ying GG, Kookana RS, Ru YJ. Occurrence and fate of hormone steroids in the environment. Environment International. 2002;**28**:545-551

[3] Thorpe KL, Cummings RI, Hutchinson TH, Scholze M, Brighty G, Sumpter JP, Tyler CR. Relative potencies and combination effects of steroidal estrogens in fish. Environmental Science & Technology. 2003;**37**:1142-1149

[4] Kostich MS, Lazorchak JM. Risks to aquatic organisms posed by human pharmaceutical use. Science of the Total Environment. 2008;**389**(2-3):329-339

[5] Desbrow C, Routledge EJ, Brighty GC, Sumpter JP, Waldock M. Identification of Estrogenic Chemicals in STW Effluent. 1. Chemical fractionation and in vitro biological screening. Environmental Science & Technology. 1998;**32**:1559-1565

[6] Forrez I, Carballaa M, Noppe H, De Brabander H, Boon N, Verstraetea W. Influence of manganese and ammonium oxidation on the removal of 17α-ethinylestradiol (EE2). Water Research. 2009;**43**:77-86

[7] De Wit M, Keil D, Van der Ven K, Vandamme S, Witters E, De Coen W. An integrated transcriptomic and proteomic approach characterizing estrogenic and metabolic effects of 17 α-ethinylestradiol in zebrafish (*Danio rerio*). General and Comparative Endocrinology. 2010; **167**:190-201

[8] Combalbert S, Hernandez-Raquet G. Occurrence, fate, and biodegradation of estrogens in sewage and manure. Applied Microbiology and Biotechnology. 2010;**86**(6):1671-1692

[9] Kolpin DW, Furlong ET, Meyer MT, Thurman EM, Zaugg SD, Barber LB, Buxton HT. Pharmaceuticals, Hormones, and Other Organic Wastewater Contaminants in U.S. Streams, 1999-2000: A National Reconnaissance. Environmental Science & Technology. 2002;**36**(6):1202-1211

[10] Washington State Department of Natural Resources and Parks. Survey of Endocrine Disruptors in King County Surface Waters. Seattle, WA: DNRP Water and Land Resources Division; 2007. p. 164

[11] Ternes TA, Stumpf M, Mueller J, Haberer K, Wilken RD, Servos M. Behavior and occurrence of estrogens in municipal sewage treatment plants - I. Investigations in Germany, Canada and Brazil. Science of the Total Environment. 1999;**225**(1-2):81-90

[12] Soares J, Coimbra AM, Reis-Henriques MA, Monteiro NM, Vieira MN, Oliveira JMA, Guedes-Dias P, Fontaínhas-Fernandes A, Parra SS, Carvalho AP, Castro LFC, Santos MM. Disruption of zebrafish (*Danio rerio*) embryonic development after full life-cycle parental exposure to low levels of ethinylestradiol. Aquatic Toxicology. 2009;**95**(4):330-338

[13] Brown KH, Schultz IR, Nagler JJ. Reduced embryonic survival in rainbow trout resulting from paternal exposure to the environmental estrogen 17α–ethynylestradiol during late sexual maturation. Reproduction. 2007;**135**:659-666

[14] Woodling JD, Lopez EM, Maldonado TA, Norris DO, Intersex VAM. other reproductive disruption of fish in wastewater effluent dominated Colorado streams. Comparative Biochemistry and Physiology, Part C: Toxicology & Pharmacology. 2006;**144**:10-15

[15] Brown KH, Schultz IR, Cloud JG, Nagler JJ. Aneuploid sperm formation in rainbow trout exposed to the environmental estrogen 17 α-ethynylestradiol. PNAS USA. 2008;**105**(50):19786-19791

[16] Martyniuk CJ, Gerrie ER, Popesku JT, Ekker M, Trudeau VL. Microarray analysis in the zebrafish (*Danio rerio*) liver and telencephalon after exposure to low concentration of 17 α-ethinylestradiol. Aquatic Toxicology. 2007;**84**:38-49

[17] Chivers AM. Investigating the Effects of 17α-Ethynylestradiol on Mitochondrial Genome Stability [thesis]. Portland: Portland State University; 2016

[18] Van den Belt K, Verheyen R, Witters H. Effects of 17 α-ethynyestradiol in a partial life-cycle test with zebrafish (*Danio rerio*): effects on growth, gonads, and female reproductive success. The Science of the Total Environment. 2003;**309**:127-137

[19] Shäfers C, Teigeler M, Wenzel A. Concentration- and Time-dependent Effects of the Synthetic Estrogen, 17 α-ethinylestradiol, on Reproductive Capabilities of the Zebrafish, *Danio rerio*. Journal of Toxicology and Environmental Health, Part A. 2007;**70**:768-779

[20] Hill RL, Janz DM. Developmental estrogenic exposure in zebrafish (*Danio rerio*): I. Effects on sex ratio and breeding success. Aquatic Toxicology. 2002;**63**:417-429

[21] Fenske M, Maack G, Schafers C, Segner H. An environmentally relevant concentration of estrogen induces arrest of male gonad development in zebrafish, *Danio rerio*. Environmental Toxicology and Chemistry. 2004;**24**(5):1088-1098

[22] Örn S, Yamani S, Norrgren L. Comparison of Vitellogenin Induction, Sex Ratio, and Gonad Morphology Between Zebrafish and Japanese Medaka After Exposure to 17 α-Ethinylestradiol and 17b-Trenbolone. Environmental Contamination and Toxicology. 2006;**51**:237-243

[23] Ortiz-Zarragoitia M, Cajaraville MP. Effects of selected xenoestrogens on liver peroxisomes, vitellogenin levels and spermatogenic cell proliferation in male zebrafish. Comparative Biochemistry and Physiology, Part C. 2005;**141**:133-144

[24] Nasiadka A, Clark MD. Zebrafish Breeding in the Laboratory Environment. ILAR Journal. 2012;**53**(2):161-168

[25] Brown KH, Dobrinski KP, Lee AS, Gokcumen O, Mills RE, Shi X, Chong WWS, Chen JYH, Yoo P, David S, Peterson SM, Raj T, Choy KW, Stranger B, Williamson RE, Zon LI, Freeman JL, Lee C. Extensive genetic diversity and strain sub-structuring among zebrafish strains revealed through copy number variant analysis. Proceedings of the National Academy of Sciences. 2012;**109**:529-534

[26] Versonnen BJ, Janssen CR. Xenoestrogenic Effects of Ethinylestradiol in Zebrafish *(Danio rerio)*. Environmental Toxicology. 2004;**19**(3):198-206

[27] Kidd KA, Blanchfield PJ, Mills KH, Palace VP, Evans RE, Lazorchak JM, Flick RW. Collapse of a fish population after exposure to a synthetic estrogen. PNAS. 2007;**104**(21):8897-8901

[28] Howe K, Clark MD, Torroja CF, Torrance J, Berthelot C, Muffato M, Collins JE, Humphray S, McLaren K, Matthews L, McLaren S, Sealy I, Caccamo M, Churcher C, Scott C, Barrett JC, Koch R, Rauch GJ, White S, Chow W, Kilian B, Quintais LT, Guerra-Assunção JA, Zhou Y, Gu Y, Yen J, Vogel JH, Eyre T, Redmond S, Banerjee R, Chi J, Fu B, Langley E, Maguire SF, Laird GK, Lloyd D, Kenyon E, Donaldson S, Sehra H, Almeida-King J, Loveland J, Trevanion S, Jones M, Quail M, Willey D, Hunt A, Burton J, Sims S, McLay K, Plumb B, Davis J, Clee C, Oliver K, Clark R, Riddle C, Elliot D, Threadgold G, Harden G, Ware D, Begum S, Mortimore B, Kerry G, Heath P, Phillimore B, Tracey A, Corby N, Dunn M, Johnson C, Wood J, Clark S, Pelan S, Griffiths G, Smith M, Glithero R, Howden P, Barker N, Lloyd C, Stevens C, Harley J, Holt K, Panagiotidis G, Lovell J, Beasley H, Henderson C, Gordon D, Auger K, Wright D, Collins J, Raisen C, Dyer L, Leung K, Robertson L, Ambridge K, Leongamornlert D, McGuire S, Gilderthorp R, Griffiths C, Manthravadi D, Nichol S, Barker G, Whitehead S, Kay M, Brown J, Murnane C, Gray E, Humphries M, Sycamore N, Barker D, Saunders D, Wallis J, Babbage A, Hammond S, Mashreghi-Mohammadi M, Barr L, Martin S, Wray P, Ellington A, Matthews N, Ellwood M, Woodmansey R, Clark G, Cooper J, Tromans A, Grafham D, Skuce C, Pandian R, Andrews R, Harrison E, Kimberley A, Garnett J, Fosker N, Hall R, Garner P, Kelly D, Bird C, Palmer S, Gehring I, Berger A, Dooley CM, Ersan-Ürün Z, Eser C, Geiger H, Geisler M, Karotki L, Kirn A, Konantz J, Konantz M, Oberländer M, Rudolph-Geiger S, Teucke M, Lanz C, Raddatz G, Osoegawa K, Zhu B, Rapp A, Widaa S, Langford C, Yang F, Schuster SC, Carter NP, Harrow J, Ning Z, Herrero J, Searle SM, Enright A, Geisler R, Plasterk RH, Lee C, Westerfield M, de Jong PJ, Zon LI, Postlethwait JH, Nüsslein-Volhard C, Hubbard TJ, Roest Crollius H, Rogers J, Stemple DL. The zebrafish reference genome sequence and its relationship to the human genome. Nature. 2013;**496**:498-503

5

Control of Programmed Cell Death During Zebrafish Embryonic Development

Nikolay Popgeorgiev, Benjamin Bonneau,
Julien Prudent and Germain Gillet

Abstract

Programmed cell death (PCD) is a conserved cellular process, which is essential during embryonic development, morphogenesis and tissue homeostasis. PCD participates in the elimination of unwanted or potentially harmful cells, and contributes in this way to the precise shaping of the developing embryo. In this review, the current knowledge related to the role of PCD during zebrafish development is described and an overview is provided about the main actors that induce, control and execute the apoptotic pathways during zebrafish development. Finally, we point out some important issues regarding the regulation of apoptosis during the early stages of zebrafish development.

Keywords: Bcl-2 family, apoptosis, cell death, embryonic development, mitochondria

1. Introduction

What would today be called genuine apoptotic cells were first observed by German scientist Carl Vogt in 1842. He was studying the morphogenesis of the tadpole notochord of the midwife toad *Alytes obstetricans* when he observed the formation and subsequent disappearance of vesicular nuclei of the embryonic notochord cells. The idea that cell death could be a fundamental inherited process was first proposed more than century later by Lockshin and Williams. They proposed that rather than a sporadic event, cell death appears in defined spatiotemporal windows during development [1]. In 1972, Kerr et al. used the term "apoptosis" meaning "to fall away from" (apo = from, ptosis = a fall), previously used to describe the falling of leaves in autumn to describe a relatively conserved set of morphological features observed in a wide variety of cell types during physiological episodes of cell death [2]. About 12 different types of programmed cell death (PCD) have been described to date, depending on the morphological features and the molecular pathways that lead to the execution of the PCD. Apoptosis, also called programmed cell death type I, is an inherited metabolically active process during which the cell dies without induction of inflammatory response. Cells undergoing apoptosis exhibit typical morphological features. Apoptotic cells appear to be shrunken and rounded shaped, without any more pseudopodia like cytoplasmic extensions. At the cytoplasmic level, mitochondria undergo fragmentation with a concomitant loss of their transmembrane potential ($\Delta\Psi m$) [3, 4]. At the nuclear level, apoptosis is characterized by typical chromatin condensation and fragmentation giving rise to pyknotic nuclei, which can be easily observed using specific dyes such as DAPI or Hoescht. Chromatin fragmentation appears to be induced by intracellular endonucleases such as caspase-activated deoxyribonuclease (CAD) and Endonuclease G (EndoG) which preferentially cut DNA strands between nucleosomes resulting in the typical "ladder pattern" observed by electrophoresis [5].

An important feature of apoptosis is the absence of inflammatory response. Indeed, the apoptotic cell maintains its plasma membrane integrity during the whole cell death process, thus preventing the intracellular proteins to interact with surrounding cells. However, phosphatidylserine (PS), an anionic phospholipid usually found at the inner leaflet of the bilayer, is exposed outwardly in apoptotic cells [6]. This morphological feature allows macrophages to detect these cells via specific PS receptors, which is then followed by rapid internalization and phagocytosis.

In vertebrates, there are two main molecular cascades for apoptosis induction [7]. The first one, called the extrinsic pathway, activates cell death by the transduction of external death signals through plasma membrane death receptors. The second one is called the intrinsic (mitochondrial) apoptosis pathway, which essentially leads to the mitochondrial outer membrane permeabilization (MOMP) and the release of apoptotic agents. Although presented at first as individual pathways, these cascades are actually interconnected. Here, we will describe the current knowledge related to the role of the apoptosis during zebrafish development.

2. The apoptotic machinery of the zebrafish embryo

2.1. Caspases

Caspases (for Cysteine ASPartate proteASE) are intracellular cysteine proteases belonging to the family of the interleukine-1β converting enzymes (ICE) family of proteases [8]. Members of the caspase family, share similar 3D conformation and are synthesized as inactive precursors called zymogens (or pro-caspases) containing a prodomain, composed of a p20 large subunit and a p10 small subunit. Caspase activation is achieved by proteolytic cleavage between the large and small subunits and removal of the N-terminus prodomain. This post-translational modification leads to a new conformational state in which caspase homodimers are fully active. The p20 subunit contains the active site of the enzyme harboring a 'QACXG' pentapeptide motif [9]. Although caspases are primarily cytosolic, they can also be found at mitochondrial and endoplasmic reticulum (ER) membranes. Caspases have been divided into three groups: interleukin activating caspases and two additional subgroups participating in the initiation and the execution of the apoptosis, respectively.

To date, 17 caspases family members have been identified in zebrafish, including initiator and effector caspases (**Table 1**). Initiator Caspases (2, 8, 9 and 10) are characterized by the presence of a long N-terminal prodomain. Zebrafish genome contains one ortholog for each of the three Caspases 2, 9 and 10, genes with conserved synteny with the human genome [10]. It also encodes for three Caspase 8 homologous genes. Caspases 8 and 10 possess two death effector domain (DED) domains in the N-terminus whereas Caspases 2 and 9 contain one caspase recruitment domain (CARD). These domains interact with adaptor proteins and are crucial for caspase activation. Importantly, Caspase 8 was identified as an actor of the death receptor pathway (cf. Section 2.3) whereas Caspase 9 belongs to the mitochondrial apoptosis pathway. In zebrafish, Caspases 2 and 8 have been identified as important regulators of vascular development. Indeed, silencing of tumor necrosis factor receptor superfamily member 1A (TNFRSF1A) expression using morpholinos led to the aberrant activation of Caspases 2 and 8 (but not of Caspase 3), resulting in apoptosis of endothelial cells [11]. Furthermore, Caspase 8 was found to be a downstream effector of the Yaf2 apoptosis regulator, a zinc finger-containing protein shown to interact with the DED domain of Caspase 8 and to inhibit apoptosis. Injection of zebrafish embryos with *yaf2* targeting morpholino did not affect gastrulation but compromised somitogenesis, which could be rescued by inhibiting Caspase 8 [12].

Effector caspases are characterized by a short N-terminus end devoid of a recruitment domain. They are activated *via* proteolytic cleavage by initiator caspases. Among effector caspases, Caspase 3 is critical for the execution of apoptosis, being activated by both, Caspases 8 and 9. In addition, Caspase 3 is capable of feed-back self-activation, thus accelerating apoptosis. Caspase 3, by targeting a wide range of vital cellular components, behaves actually as a genuine apoptosis executor. Activation of Caspase 3 is often considered as a no-return point of apoptosis. Among the substrates cleaved by Caspase 3, are found cytoskeleton proteins, anti-apoptotic factors, metabolic enzymes and several nucleases. For example, proteolytic cleavage of the inhibitory domain of CAD endonuclease (ICAD) leads to CAD activation and subsequent DNA fragmentation, a typical feature of the apoptotic process. The zebrafish genome encodes two Caspase 3 homologs, namely Caspases 3a and 3b [13, 14]. In fact, Caspase 3 activity has been first assessed in stressful conditions following cycloheximide or staurosporine treatments at mid-gastrula stage. In these conditions embryonic development is rapidly blocked, with Caspase 3 being activated within 8 h [15]. Furthermore, Yamashita and colleagues generated a transgenic zebrafish strain expressing full length pro-Caspase 3 to study the role of Caspase 3 in embryonic development. Indeed, these transgenics exhibited a marked increase in the number of apoptotic cells specifically in the retina, the notochord, the heart and the yolk sac, suggesting an essential role of Caspase 3 in numerous morphogenetic processes. Interestingly, silencing of *caspase 3* using specific morpholinos did not lead to any significant developmental defects, suggesting some redundancy with other effector caspases, such as Caspases 6 and 7 [16].

In addition, the zebrafish genome encodes for two caspases belonging to the interleukin activating caspases. These caspases known as Caspy (Caspy and Caspy2) contain N-terminal pyrin domains [17]. In the case of Caspy, the pyrin domain was found to be essential for its interaction with the apoptosis-associated speck-like protein containing a CARD (zAsc). In effect, zAsc binding to Caspy led to its activation and apoptosis execution *in cellulo*. In zebrafish, both genes are specifically expressed in the pharyngeal arches, *caspy* silencing resulting in developmental defects in this particular region.

Type	Mammalian protein	Zebrafish homologs	Accession number
Death receptor ligands	TNF (TNFSF2)	Tnfa (Tnf1)	NM_212859
		Tnfb (Tnf2)	NM_00102444
	CD95/FasL (TNFSF6)	Faslg (Fas ligand)	NM_001042701
	Apo2L/TRAIL (TNFSF10)	Tnfsf10 (Tnfsf10l2)	NM_001002593
		Tnfsf10l (DL1a)	NM_131843
		Tnfsf10l3 (DL1b)	NM_001042713
		Tnfsf10l4 (DL3)	NM_001013283
	APP	Appa	NM_131564
	TL1A	si:ch211-158d24.4 (tnfsf15)	NM_001123259
Death receptor	TNFR1 (TNFRSF1A)	Tnfrsf1a	NM_213190
	CD95/Fas (TNFRSF6)	Fas	XM_021467407
	TNFRSF10A and B (DR4 and DR5)	Hdr (ZH-DR)	NM_194391
	TNFRSF10A and B (DR4 and DR5)	Tnfrsfa (OTR)	NM_131840
	TNFRSF21 (DR6)	Tnfrsf21 (DR6)	NM_001042688
Adaptor protein	FADD	Fadd	XM_001923858
Initiator caspases	Caspase 2		NM_001042695
	Caspase 4	Caspy	NM_131505
		Caspy 2	NM_152884
		Caspase b, like	NM_001145592
		zgc:171731	NM_001109712
	Caspase 8	Caspase 8a	NM_131510
		Caspase 8 l1	NM_001098619
		Caspase 8 l2	XM_680338
Effector caspases	Caspase 3	Caspase 3a	NM_131877
		Caspase 3b	NM_001048066
	Caspase 6	Caspase 6a	NM_001020497
		Caspase 6b	NM_001005973
		Caspase 6c	NM_001039980
	Caspase 7	Caspase 7a	NM_001020607
Inhibitor of caspases	c-FLIP (c-FLAR)	Cflara (Cflar)	NM_001313772
	Birc 2	Birc 2, IAP 1	NM_194395
	Birc 4	Birc 4, XIAP	NM_194396

Type	Mammalian protein	Zebrafish homologs	Accession number
	Birc 5	Birc 5a, Survivin	NM_194397
		Birc 5b, Survivin 2	NM_145195
	Birc 6	Birc 6	XM_009293036.3
	Birc 7	Birc 7	NM_001098768
	Hsp 70	Hsp70	AB062116.1
BCL-2 family	Bcl-xL	zBlp1	NM_131807
(BCL-2 like)	Bcl-2	zBlp2	NM_001030253
	Mcl-1	Mcl-1a	NM_131599
		Mcl-1b	NM_194394
	Nrh, Bcl-2 l10	Nrz	NM_194398
(BAX-like)	Bax	Baxa, zBax1	NM_131562
		Baxb, Zbax2	NM_001013296
	Bok	Boka	NM_001003612
		Bokb	NM_201185
	—	Bcl-wav, Bcl-2 l16	NM_001172402
	Bcl-2 l13, Bcl-rambo	Bcl-2 l13	NM_001044891
(BH3-only)	Bad	Bada	XM_005161364
		Badb	NM_001270595
	Bbc3, Puma	zPuma, Bbc3	NM_001045472
	Noxa, Pmaip1	zNoxa, Pmaip1	NM_001045474
	Bim	zBim, Bcl2-l11	NM_001135791
	Bid	zBid	NM_001079826
	Bik	zBik	NM_001045038
	Bmf	zBmf1	NM_001045224
		zBmf2	NM_001045473
	Bnip1	Bnip1a	XM_684156
		Bnip1b	XM_001333689
	Bnip2	Bnip2	NM_201218
	Bnip3l	Bnip3la, Nix	NM_001012242
		Bnip3lb, Nip3a	NM_205571
	Bnip4l	Bnip4l, Nip3b	NM_212693
		Bnipl	NM_001128394

Type	Mammalian protein	Zebrafish homologs	Accession number
MIRAF*	Cytochrome C	Cytochorome C, Cycsb	NM_001002068
	EndoG	EndoG	NM_001024214.1
	AIF	AIF, Aifm1	NM_200102.2
	Smac/Diablo	Diabloa	NM_200346
		Diablob	NM_001243034
	HtrA2/Omi	LOC110437853	XM_021472675.1
Others	Apaf-1	Apaf-1	NM_001045243
	P53	p53, TP53	NM_131327

Accession numbers from NCBI database were presented on the left.
*MIRAF, mitochondria released apoptotic factors.

Table 1. Summary table of apoptosis-associated genes found in the zebrafish genome.

2.2. Caspase inhibitors

Since the discovery of viral caspase inhibitors, it became clear that multicellular organisms were also able to make their own caspase inhibitors. These proteins called inhibitors of apoptosis proteins (IAPs) are characterized by one or more baculoviral IAP repeats (BIR), allowing them to prevent caspase activation and apoptosis. The IAPs can also possess a RING domain with an ubiquitin-ligase activity at their C-terminus end. This feature allows the IAPs not only to block caspase activity but also to promote their degradation by the proteasome [18]. It should be noted that six IAP proteins are found in zebrafish [19]. Zebrafish IAPs including survivins (BIRC5a and BIRC5b) appear to play a role in embryonic development. Indeed, the knockdown of *birc5a* leads to multiple defects including in the nervous system, the cardiovascular system and the hematopoietic system. Interestingly, this phenotype was rescued by ectopic expression of both *birc5* paralogs, suggesting the existence of functional redundancy during embryogenesis [20, 21].

Another well characterized caspase inhibitor is the Hsp70 chaperone. Actually, under stressful conditions, the cell can protect itself from the uncontrolled activation of the apoptosis by increasing Hsp70 levels. This protein can bind to and block the recruitment of initiator Caspase 9 into the apoptosome complex [22]. In zebrafish, injection of *hsp70* targeting morpholinos resulted in "small eye" phenotype. Close analysis of this phenotype identified a significant increase in apoptotic cells specifically in the developing lens of the zebrafish embryo [23, 24].

Cellular FLICE-like inhibitory protein (c-FLIP) is another example of a cellular caspase inhibitor. This structural analog of Caspase 8 is devoided of proteolytic activity and, is able to bind to DD and prevents the activation of Caspase 8 downstream of the death receptors pathway [25]. c-FLIP is specific to the vertebrate lineage. The knockdown of *c-flip* in zebrafish results in important cardiovascular abnormalities, including cardiac edema and irregular blood flow consecutive to the formation of blood clots in the vessels [26].

2.3. Molecular actors linked to the death receptor pathway

The cell-extrinsic (or death receptor) pathway of apoptosis is activated by the binding of extracellular ligand proteins belonging to the tumor necrosis factor (TNF) superfamily to specialized receptors, the death receptors called TNF receptors [27]. In mammals, six death receptors have been characterized together with five "death-inducing" ligands. The death receptors are characterized by at least one extracellular cystein-rich domain (CRD) allowing the recognition between the ligand and the receptor and by an intracellular conserved domain called the death domain (DD). Activation of death receptor pathway induces clustering of the receptors through their pre ligand-binding assembly domain (PLAD) [28]. This clustering triggers the recruitment of adaptor proteins such as Fas-associated protein with death domain (FADD) which interact with the DD of the receptor but also with initiator caspases (Caspases 8 and 10) thus forming the death-inducing signaling complex (DISC) [29]. DISC formation then induces activation of initiator caspases leading to the activation of effector caspases and apoptosis execution [30]. DISC-dependent caspases activation can be inhibited by c-FLIP [25].

In zebrafish, on the basis of phylogenetic analysis, orthologs of the five death receptors ligands have been identified (**Table 1**). CD95/FasL, APP and TL1A each possess one zebrafish ortholog (*faslg*, *appa* and *158d24.4*, respectively) while TNF possesses two orthologs (TNFa and TNFb) and Apo2L/TRAIL five of them (TNFSF10L, TNFSF10L2, TNFSF10L3 and TNFSF10L4) [31, 32].

The genes encoding death receptors have also been identified in zebrafish. Based on the presence and organization of their CRD and DD, clear orthologs of TNFR1 (TNFRSF1A), CD95/Fas (Fas) and DR6 (TNFRSF21) were characterized. Notably, a selective interaction between TNFa and TNFRSF1 were confirmed by immunoprecipitations [31]. Two other DD-containing receptors, HDR and TNFRSFA, have been described in zebrafish [33, 34]. Extracellular domains of these receptors are close to the one of CD95/Fas but their DD are more similar to DR4 and DR5 DD. However, the observation that HDR and TNFRSFA both bind three orthologs of Apo2L/TRAIL (TNFSF10L, TTNFSF10L2 and TTNFSF10L3) and that they are required for apoptosis induced by these ligands in zebrafish embryos strongly suggests that HDR and TNFRSFA are in fact orthologs of DR4 and DR5 [31]. Finally, so far, no zebrafish ortholog of DR3 have been characterized but the existence of an ortholog of TL1A suggests that a zebrafish DR3 will be identified shortly.

In addition to death receptors and their ligands, the components of DISC are also conserved in zebrafish. A clear ortholog of FADD, containing a DD and a death effector domain (DED), has been identified as well as an ortholog of Caspase 8, *casp8a* [31, 35]. The latter possesses an N-terminal DED which allow association with FADD and a QACQG active-site motif that is characteristic of Caspases 8 and 10 in mammals. Caspase 8a and FADD are both required for apoptosis induced by Apo2L/TRAIL orthologs in zebrafish embryos. Moreover, Caspase 8a is functionally conserved as it restores death receptor-induced apoptosis in mouse cells lacking endogenous Caspase 8 [35]. Two other genes related to *casp8a* exist (*casp8l1* and *casp8l2*) but their possible involvement in the cell-extrinsic pathway is unclear. Indeed, Caspase 8l1 has a QACQG active-site motif but no DED whereas Caspase 8l2 possesses an N-terminal DED but its active site is similar to the one of Caspase 2. Finally, a zebrafish ortholog of c-FLIP, referred to as Cflar, able to inhibit apoptosis induced by Apo2L/TRAIL orthologs have also been described [31].

In mammals, the cell-extrinsic apoptosis pathway is essential for the functioning of the immune system. However, this has not been clearly established in zebrafish as yet. One study revealed that apoptosis is important for T and B cells homeostasis as overexpression of Bcl-2 in these cells increased their number, but this work did not show a role for the extrinsic pathway in lymphocyte homeostasis [36]. In contrast, the death receptor HDR appears to be involved in red blood cells homeostasis. Indeed, HDR is specifically expressed in hematopoietic lineage and its inhibition, using either a dominant negative mutant or antisense morpholinos, leads to abnormal accumulation of erythroid cells [34, 37].

During early development, *hdr*, *tnfrsfa* and *fadd* are expressed in the notochord suggesting that they might play a role in this structure [31]. Fas and FasLg are also present in developing notochord and their knockdown *via* morpholinos leads to prolonged expression of notochord specific genes and to an abnormally enlarged notochord at 4 days post-fertilization (dpf). As apoptosis plays a role in notochord regression, this suggests that extrinsic pathway may be involved in this process [38].

During development, TNFRSFA and its ligands TNFSF10L2 and TNFSF10L3 are expressed in particular in neuromasts that contain hair cells which turnover is regulated by apoptosis [31, 39]. Finally, the death receptor pathway seems to be involved in zebrafish eye development as FADD is required for slowing cell growth during this process [40].

2.4. Molecular actors linked to the intrinsic pathway

The intrinsic pathway of apoptosis also called the mitochondrial pathway can be induced by various signals including DNA damage, chemotherapy, viral infection or growth factors deprivation. The mitochondrial pathway of apoptosis is mainly controlled by the Bcl-2 family of proteins, which is described in more detail in the next chapter. The mitochondrion is a membrane-enclosed organelle found in most eukaryotic cells [41]. The diameter of these organelles falls under the 0.5–10 µm range. Mitochondria are often referred as the "power-house" of the cell because they generate most of the cell's supply of adenosine triphosphate (ATP). In addition to their bioenergetic function, mitochondria are involved in a number of other processes, such as signal transduction, cell differentiation, cell cycle and cell growth [42]. Mitochondria are also major "decision centers" for the execution or prevention of apoptosis. Indeed, a number of pro-apoptotic molecules appear to be stored in the existing space between the inner mitochondrial membrane (IMM) and the OMM. At the mitochondrial level, the induction of apoptosis leads to the OMM permeabilization, which leads to the irreversible release into the cytosol of pro-apoptotic factors that promote caspase activation, DNA fragmentation and ultimately the death of the cell. Belong to this toxic molecular "cocktail" among others: Cytochrome C, AIF, EndoG, Smac/Diablo and HtrA2/Omi.

2.4.1. Mitochondria released apoptotic factors

Cytochrome C is a small heme protein (approximately 12 kDa), exhibiting a positive net charge, located in the intermembrane space (IMS) where it can be loosely attached to the IMM. Cytochrome C is synthesized inside the cytosol, and subsequently transported into the IMS. Bioenergetically, Cytochrome C is a component of the mitochondrial electron transport chain. The heme molecule of cytochrome C accepts electrons originating from complex III and transfers them to the cytochrome oxydase complex, thus cytochrome C is indispensable for the oxidative phosphorylation and the maintenance of cellular energy fluxes. Cytochrome C plays an additional role in the context of apoptosis, as it is now well established by a large number of *in vitro* and *in vivo* studies [43–46]. The release of Cytochrome C is a rapid and complete process, irrespective of the intensity of the death-inducing signal or the temperature, indicating that this is a non-enzymatic phenomenon obeying an "all or nothing" law. When released in the cytosol, Cytochrome C interacts with the adaptor protein apoptosis protease activating factor (Apaf-1) in presence of ATP. Apaf-1 was first characterized by Wang and collaborators [47]. It is a 130 kDa multidomain protein, containing a CARD domain at its N-terminus end, sharing homology with the CARD domain of Caspase 9, an ATPase domain as well as two WD-40 repeats at the C-terminus end. The Wang laboratory discovered the existence of a multiprotein complex called the apoptosome, comprising Cytochrome C, Apaf-1 and Caspase 9, which was found to be able to activate Caspase 3 [48]. The apoptosome was crystallized by Acehan and collaborators in 2002 [49]. Although the existence of the zebrafish apoptosome has not been directly demonstrated, its genome encodes for all functional components of this complex. Furthermore, treatment with drugs converging toward the intrinsic pathway was found to lead to Caspase 3 activation [50, 51].

Apoptosis-inducing factor (AIF) is a 57 kDa flavoprotein with NADH oxidase activity which is located in the mitochondrial IMS [52, 53]. As Cytochrome C, AIF is encoded by a nuclear gene, which is imported into the mitochondria after being synthesized in the cytosol. In response to diverse death stimuli, AIF is released into the cytosol and transferred to the nucleus where it binds to chromatin [54]. The binding of AIF to DNA induces chromatin peripheral condensation and subsequent fragmentation. Wang and colleagues showed that this process is due to AIF-dependent activation of endogenous nucleases such as endonuclease G and CAD. AIF is unable to induce the fragmentation of the DNA on its own. To this end, AIF needs the help of endonuclease G. EndoG is a mitochondrial nuclease of 30 kDa which is required for the replication of the mitochondrial chromosome. During apoptosis, EndoG is released from the mitochondria into the cytosol and subsequently enters into the nucleus. In this compartment, EndoG cleaves the DNA into nucleosomal fragments [55]. AIF and EndoG orthologs are both expressed during zebrafish embryonic development but their functional implications remain to be analyzed. Second mitochondria-derived activator of caspase/direct IAP-binding protein with low pI (Smac/Diablo) is a mitochondrial protein, which resides in the intermembrane space [56, 57]. The human gene is called *smac* whereas the mouse ortholog is called *diablo*. Smac/Diablo is synthesized as a precursor protein containing a 55 residue N-terminal mitochondrial targeting sequence. During the translocation of Smac/Diablo to the mitochondria, this sequence is cleaved which uncovers an IAP-binding motif (IBM) required for apoptotic activity. Indeed, in the presence of an apoptotic stress, the mitochondria release Smac/Diablo in the cytosol where it participates indirectly in the activation of caspases by binding IAP proteins. The binding of Smac/Diablo to IAP disrupts IAP-Caspase interactions. Thus, released caspases can then be activated and execute the cell death program. The Smac/Diablo-IAP complex formation appears to be a regulated process since IAP can ubiquitinate Smac/Diablo and drive it for proteasomal degradation [18]. In zebrafish, Smac/Diablo gene is mainly expressed in the late developmental stages with the most prominent expression in the heart, the lens and the liver. However, the possible implication of this IMS factor in the morphogenesis of these organs remains unknown.

High temperature requirement A2 (HtrA2/Omi) is a heat shock protein first identified in *Escherichia coli* [58]. Its ortholog in mammals, called Omi was initially described as an ER protein

[59, 60]. However, it seems that this protein is mainly mitochondrial in non-apoptotic cells [61]. When the OMM is permeabilized, Omi is released into the cytosol where it binds to IAP and irreversibly inhibits its activity by proteolytic cleavage [57]. Thus, it seems that Omi is a more potent inhibitor of IAP than Smac/Diablo which blocks IAP reversibly. Currently, there is no data about the possible implication of HtrA2/Omi functional implication in zebrafish have not been assessed.

2.4.2. Bcl-2 family of proteins

The Bcl-2 family of proteins is a group of intracellular eukaryotic proteins best known for their implication in MOMP. The founding member of this family, the *bcl2* (B-cell lymphoma/leukemia 2) gene was discovered in a study on a chromosome translocation frequently observed in human B-cell follicular lymphomas. Tsujimoto and colleagues showed that the translocation between chromosome 18 (q21) chromosome 14 (q32), t (14;18), results in the relocation of the *bcl2* ORF downstream of the enhancer promoter region of the *igh* heavy chain immunoglobulin gene [62–64]. This translocation results in transcriptional upregulation of *bcl2* gene expression [65]. The Bcl-2 family comprises proteins with antagonistic functions with respect to apoptosis regulation. Structurally, all Bcl-2 proteins contain in their primary structure one or more conserved Bcl-2 homology (BH) domains. Based on this criterion, three subgroups have been identified: (1) the anti-apoptotic multidomain members containing all four BH domains (BH1–4); (2) the pro-apoptotic members containing three BH domain (lacking BH4) and (3) the proapoptotic BH3-only members containing only the sole BH3 domain. In addition, Bcl-2 proteins may contain a transmembrane anchoring domain (TM domain) at the C-terminus end. Bcl-2 multidomain pro-apoptotic members (Bax and Bak) are the effectors for MOMP. Through their oligomerization, they form pores at the MOM and promote the release of apoptotic factors including Cytochome C, AIF and EndoG [66]. Anti-apoptotic Bcl-2 members block their activity and promote cell survival, whereas BH3-only proteins play the role as intracellular judges as they can inhibit anti-apoptotic members and/or promote Bax/Bak oligomerization.

Bcl-2 homologs of all three subgroup of Bcl-2 family have been identified and molecularly characterized in zebrafish [67–69]. Due to genomic duplication, the zebrafish genome possesses several Bcl-2-related paralogs including Mcl-1a, Mcl-1b, Bax1, Bax2, Boka, Bokb and Bmf1 and Bmf2. Interestingly, the ortholog of the *bak* gene has been lost in the teleost lineage. Instead the zebrafish genome harbors *bcl-wav* (an acronym for Bcl-2 homolog found in water living anamniotes), a *bcl-2* homolog only found in fishes and anurans [70]. With the exception of the BH3-only protein Bik, all Bcl-2 related genes in zebrafish are maternally inherited since their corresponding mRNAs were detected before the mid-blastula transition [67].

Of note during zebrafish early development physiological apoptosis is not observed suggesting that at least some of the Bcl-2 family members may have additional non-apoptotic roles during early embryogenesis. In this respect, Zhong and colleagues recently demonstrated that the zebrafish ortholog of the BH3-only protein Noxa (zNoxa) not only controls apoptosis during late stages of gastrulation but also plays role in cell cycle in the developing blastula [71]. Indeed *znoxa* knockdown led to a significant decrease in the number of mitotic cells. This phenotype seems to be dependent on the Wnt signaling pathway since *znoxa* knockdown led to increase of *zwnt4b* expression. In addition silencing of *zwnt4b* rescued zNoxa phenotype and restored a WT count of cells in G2/M phase.

Using the zebrafish model, we demonstrated that a Bcl-2 homolog, referred to as Nrz (Nr-13 ortholog in zebrafish), is critical during the early stages of zebrafish development [72]. In zebrafish, Nrz protein possesses a dual subcellular localization at the ER and the mitochondria [73]. Its functional invalidation causes embryo development arrest followed by detachment of the entire blastomeres from the yolk sac. By performing a series of time-lapse and confocal microscopy experiments, we demonstrated that this phenotype is due to the premature formation of the actin-myosin contractile ring, a supramolecular structure, which squeezes and halves the embryo at the level of the margin. Furthermore, by using single subcellular localization (SSL) Nrz mutants, we showed that the ER-resident Nrz but not mitochondrial-resident Nrz was critical for its physiological function. Indeed, at the ER membrane, Nrz interacts with the Ca^{2+} channel inositol trisphosphate receptor type 1 (IP_3R1) *via* its BH4 domain [73, 74]. In this way, Nrz slows down the release of Ca^{2+} into the yolk sac, which consecutively allows controlling the formation of the contractile actin-myosin ring *via* the Calmodulin-MLCK pathway. Overall, our results highlighted for the first time that a Bcl-2 family member is able to orchestrate cellular migration events by controlling intracellular Ca^{2+} fluxes.

In addition, we identified the new Bcl-2 family member Bcl-wav [75]. The *bclwav* gene is expressed throughout zebrafish early development; it encodes a pro-apoptotic Bcl-2 family member with strict mitochondrial localization. However, *bclwav* silencing in zebrafish causes a specific apoptosis independent phenotype at 24 hpf. Macroscopically, this phenotype is characterized by an embryo anterioposterior axis reduction as well as notochord deviation. Using time-lapse microscopy, we demonstrated that this phenotype affects the convergence and extension movements which underlie the establishment of the embryonic axes. Indeed, in *bclwav*-silenced embryos, mesodermal cells migrated erratically compared to cells from control embryos, which moved in a coordinated fashion. *In vivo* analysis of the actin cytoskeleton revealed that these migration defaults correlated with randomization of F-actin protrusion dynamics. Interestingly, *bclwav* silencing was correlated with a decrease of mitochondrial Ca^{2+} levels and concomitant increase of cytosolic Ca^{2+}. Together these results indicated that Bcl-wav controls the dynamics of the actin cytoskeleton by regulating intracellular Ca^{2+} homeostasis at the mitochondrial level. Indeed, at the mitochondria, Bcl-wav interacts with the voltage-dependent anion channel (VDAC) channel and enhances mitochondrial Ca^{2+} uptake, which in turn controls actin polymerization and cell migration. It is important to note that, *mitochondrial calcium uniporter* (*mcu*) knockdown phenocopies *bclwav*-silenced embryos. Indeed, MCU downregulation leads to decreased mitochondrial calcium uptake and impaired actin dynamics giving the first insights into the critical role of the mitochondrial Ca^{2+} oscillations in the vertebrate development [70].

Put into a broader context, the results demonstrate that members of the Bcl-2 family are able to control cell migration in a calcium-dependent manner *via* their direct interaction with intracellular Ca^{2+} channels independently of their involvement in the regulation of cell death [76].

2.4.3. P53

P53 is a transcription factor considered as the main tumor suppressor regulating cell fate decisions. Indeed, p53 is the most frequently mutated and/or inactivated gene in human cancer modulating cell responses to DNA damage, oncogenic signaling and hypoxia in order to preserve genome integrity. The zebrafish p53 ortholog is highly conserved with 48% of sequence identity compared to the human sequence [77]. P53 is highly and ubiquitously expressed during early zebrafish embryo development, then predominantly expressed in the brain during the pharyngula stage before to decrease in expression after 48 hpf [77–79]. P53 tumor suppressor activity has been confirmed in the zebrafish model by the characterization of a p53 mutant harboring a missense mutation in the DNA-binding domain, M214K, leading to the development of multiple organ tumors at around 8.5 months [80]. The mechanism of action of p53 has been widely studied in mammals and similar results have been obtained in the zebrafish model [81]. During different cell stresses, p53 expression is stabilized leading to its activation and accumulation in the nucleus, and subsequently to cell cycle arrest and the intrinsic pathway of apoptosis [82].

Its role in DNA damage and apoptosis has been well studied in the zebrafish model [82]. It has been shown that increased DNA damage leads to the stimulation of p53 transcription and an increase of p53 protein level [79]. In addition, knockdown of *p53* decreases apoptosis induced by different stimuli including gamma and UV irradiation, camptothecin treatment or altered DNA replication [83–85].

Finally, an undesirable effect of p53 activation-induced cell death has been characterized and associated with off-target effects triggered by morpholino antisense oligonucleotides technologies. These off-target effects have been assimilated to p53 signaling pathway-dependent apoptosis [86, 87]. Indeed, the most commonly off-target reported phenotype is characterized by an increase of neural cell death at 22 hpf [86]. Even if the precise molecular mechanism is still unclear, it has been attributed to the activation of p53 leading to the apoptotic cell death. Indeed, the role of p53 in this phenotype has been shown by the characterization of *smo* and *wnt5* morphant embryos, where simultaneous knockdown of p53 in these embryos rescued neuronal apoptosis [87]. These results highlight the extremely cautious, which has to be employed when potential actors of the apoptotic program are studied by the morpholino strategy. In addition to the use of different morpholino sequences targeted, the same gene and crucial rescue experiments, co-expression of p53 morpholino is now commonly used to discern between gene-specific and off-target effects [19]. Indeed, p53 is not required for proper early embryos development and p53 morpholino does not interfere with other gene-specific phenotypes [87, 88].

3. Developmental control of apoptosis in the early zebrafish embryo

Following fertilization and during early stages of embryonic development, embryo relies entirely on the maternal inherited mRNAs and proteins which were accumulated during oogenesis. After several synchronous divisions, which lack G1 and G2 phases, the cell cycle slows down and divisions become asynchronous. This step, referred to as the mid-blastula transition (MBT), corresponds to the beginning of the expression of zygotic genes. Ikegami and colleagues first noticed that zebrafish embryos treated before the MBT with microtubule destabilizing agent nocodazole or DNA-damaging molecules such as camptothecine, hydroxyurea or aphidicoline did not result in direct apoptosis activation. Instead, the cell cycle was arrested with the apoptotic program being executed several hours later, during the mid-gastrula stage [50, 51]. This phenomenon is not restricted to zebrafish as a similar apoptotic control also operates in the Xenopus embryo [89]. These observations suggested that key molecular components of the apoptotic program were either missing or inactivated during early development. In this respect, one report using a proteomics approach identified that Apaf-1 was missing before the MBT suggesting that a functional apoptosome may be set up after the MBT transition [90]. However, these later data do not explain why inhibition of protein synthesis using cycloheximide is not able to slow the apoptotic program in the zebrafish gastrula [15]. Furthermore, our laboratory demonstrated that ectopic expression of zBax through recombinant mRNA injection in one cell stage embryos actually led to $\Delta\Psi m$ loss and Caspase 3 activation as early as the blastula stage [91]. Importantly, this apoptotic phenomenon specifically occurs in an extraembryonic structure referred to as the yolk syncytial layer (YSL). The YSL results from the fusion of blastomeres physically connected with the yolk cell by cytoplasmic bridges [92]. Fusion of margin blastomeres with the yolk leads to the release of cell nuclei and other cellular components including a dense network of active mitochondria interconnected with ER membranes. Purified YSL mitochondria can undergo MOMP and Cytochrome C release. This was demonstrated by performing *in vitro* Cytochrome C release assay using recombinant human truncated Bid protein. Indeed, Bid is a BH3-only protein which once cleaved by Caspase 8 translocates to the MOM and activates Bax oligomerization. Thus it is tempting to speculate that at least at the level of the YSL mitochondria harbor sufficient amount of Bax in order to initiate MOMP following BH3-only stimulus.

Altogether these results showed that zebrafish early embryo possesses a functional apoptotic machinery. Thus the tight apoptotic control observed by Ikegami et al. may be exerted at the post-translational level through protein-protein interactions. In this respect, Kratz and colleagues demonstrated that manipulation of the ratio between pro- or anti-apoptotic Bcl-2 proteins determines the capacity of early zebrafish embryo to undergo apoptosis. Notably, overexpression of BH3-only or zBax paralogs induced rapid Caspase 3-dependent cell death whereas co-expression of Bcl-2-related anti-apoptotic members effectively counteracted early embryo mortality [67].

4. Conclusion

Apoptosis represents a key cellular process that maintains tissue homeostasis and shapes the embryo. Impairment or *a contrario* overactivation of cell death often leads to severe developmental abnormalities and lethal phenotypes. Thus, the tight spatiotemporal control over apoptosis induction is critical for orchestrating embryonic development. The fact that zebrafish genome encodes for the majority of apoptosis actors found in the human genome makes zebrafish a valuable model for understanding the contribution of apoptosis regulators during embryonic development in vertebrates. The use of antisense chemically modified nucleotides, most notably morpholinos, allowed to assess the implication of many apoptosis regulators in the developmental process. However, the possible off targeting and unspecific activation of the p53 pathway can be a drawback in some instances. In this respect, the development of new genome editing approaches such as CRISPR/Cas9 will allow in the near future to assess the precise role of each and every member of the cell death machinery during embryogenesis.

Acknowledgements

This work is supported by AFM telethon, Ligue Nationale Contre le Cancer, Cancéropole Auvergne Rhônes- Alpes (CLARA—Oncostarter), Fondation ARC and Medical Research Council, UK (MC_UU_00015/7).

Conflict of interest

The authors declare that they have no conflict of interest.

Author details

Nikolay Popgeorgiev[1]*, Benjamin Bonneau[2], Julien Prudent[3] and Germain Gillet[1,4]

*Address all correspondence to: nikolay.popgeorgiev@univ-lyon1.fr

1 Université de Lyon, Centre de recherche en cancérologie de Lyon, U1052 INSERM, UMR CNRS 5286, Université Lyon I, Centre Léon Bérard, Lyon, France

2 Institut NeuroMyoGene, Université Claude Bernard Lyon 1, Centre National de la Recherche Scientifique, Unité Mixte de Recherche 5310, Institut National de la Santé et de la Recherche Médicale U1217, Lyon, France

3 Medical Research Council Mitochondrial Biology Unit, University of Cambridge, Wellcome Trust/MRC Building, Cambridge Biomedical Campus, Hills Road, Cambridge, United Kingdom

4 Hospices civils de Lyon, Laboratoire d'anatomie et cytologie pathologiques, Centre Hospitalier Lyon Sud, chemin du Grand Revoyet, Pierre Bénite, France

References

[1] Lockshin RA, Williams CM. Programmed cell death--I. Cytology of degeneration in the intersegmental muscles of the Pernyi Silkmoth. Journal of Insect Physiology. 1965; **11**:123-133

[2] Kerr JF, Wyllie AH, Currie AR. Apoptosis: A basic biological phenomenon with wide-ranging implications in tissue kinetics. British Journal of Cancer. 1972;**26**(4):239-257

[3] Adams JW et al. Cardiomyocyte apoptosis induced by Galphaq signaling is mediated by permeability transition pore formation and activation of the mitochondrial death pathway. Circulation Research. 2000;**87**(12):1180-1187

[4] Zamzami N et al. Inhibitors of permeability transition interfere with the disruption of the mitochondrial transmembrane potential during apoptosis. FEBS Letters. 1996;**384**(1):53-57

[5] Wyllie AH. Glucocorticoid-induced thymocyte apoptosis is associated with endogenous endonuclease activation. Nature. 1980;**284**(5756):555-556

[6] Fadok VA et al. Regulation of macrophage cytokine production by phagocytosis of apoptotic and post-apoptotic cells. Biochemical Society Transactions. 1998;**26**(4):653-656

[7] Elmore S. Apoptosis: A review of programmed cell death. Toxicologic Pathology. 2007; **35**(4):495-516

[8] Alnemri ES et al. Human ICE/CED-3 protease nomenclature. Cell. 1996;**87**(2):171

[9] Fuentes-Prior P, Salvesen GS. The protein structures that shape caspase activity, specificity, activation and inhibition. The Biochemical Journal. 2004;**384**(Pt 2):201-232

[10] Sidi S et al. Chk1 suppresses a caspase-2 apoptotic response to DNA damage that bypasses p53, Bcl-2, and caspase-3. Cell. 2008;**133**(5):864-877

[11] Espin R et al. TNF receptors regulate vascular homeostasis in zebrafish through a caspase-8, caspase-2 and P53 apoptotic program that bypasses caspase-3. Disease Models & Mechanisms. 2013;**6**(2):383-396

[12] Stanton SE et al. Yaf2 inhibits caspase 8-mediated apoptosis and regulates cell survival during zebrafish embryogenesis. The Journal of Biological Chemistry. 2006;**281**(39): 28782-28793

[13] Yabu T et al. Characterization of zebrafish caspase-3 and induction of apoptosis through ceramide generation in fish fathead minnow tailbud cells and zebrafish embryo. The Biochemical Journal. 2001;**360**(Pt 1):39-47

[14] Tucker MB et al. Phage display and structural studies reveal plasticity in substrate specificity of caspase-3a from zebrafish. Protein Science. 2016;**25**(11):2076-2088

[15] Negron JF, Lockshin RA. Activation of apoptosis and caspase-3 in zebrafish early gastrulae. Developmental Dynamics. 2004;**231**(1):161-170

[16] Yamashita M et al. Extensive apoptosis and abnormal morphogenesis in pro-caspase-3 transgenic zebrafish during development. The Journal of Experimental Biology. 2008; **211**(Pt 12):1874-1881

[17] Masumoto J et al. Caspy, a zebrafish caspase, activated by ASC oligomerization is required for pharyngeal arch development. The Journal of Biological Chemistry. 2003; **278**(6):4268-4276

[18] MacFarlane M et al. Proteasome-mediated degradation of Smac during apoptosis: XIAP promotes Smac ubiquitination in vitro. The Journal of Biological Chemistry. 2002; **277**(39):36611-36616

[19] Eimon PM, Ashkenazi A. The zebrafish as a model organism for the study of apoptosis. Apoptosis. 2010;**15**(3):331-349

[20] Delvaeye M et al. Role of the 2 zebrafish survivin genes in vasculo-angiogenesis, neurogenesis, cardiogenesis and hematopoiesis. BMC Developmental Biology. 2009;**9**:25

[21] Ma A et al. The role of survivin in angiogenesis during zebrafish embryonic development. BMC Developmental Biology. 2007;7:50

[22] Beere HM et al. Heat-shock protein 70 inhibits apoptosis by preventing recruitment of procaspase-9 to the Apaf-1 apoptosome. Nature Cell Biology. 2000;2(8):469-475

[23] Evans TG et al. Heat shock factor 1 is required for constitutive Hsp70 expression and normal lens development in embryonic zebrafish. Comparative Biochemistry and Physiology. Part A, Molecular & Integrative Physiology. 2007;146(1):131-140

[24] Evans TG et al. Zebrafish Hsp70 is required for embryonic lens formation. Cell Stress & Chaperones. 2005;10(1):66-78

[25] Krueger A et al. Cellular FLICE-inhibitory protein splice variants inhibit different steps of caspase-8 activation at the CD95 death-inducing signaling complex. The Journal of Biological Chemistry. 2001;276(23):20633-20640

[26] Sakamaki K et al. Conservation of structure and function in vertebrate c-FLIP proteins despite rapid evolutionary change. Biochemistry and Biophysics Reports. 2015;3:175-189

[27] Aggarwal BB. Signalling pathways of the TNF superfamily: A double-edged sword. Nature Reviews. Immunology. 2003;3(9):745-756

[28] Chan FK et al. A domain in TNF receptors that mediates ligand-independent receptor assembly and signaling. Science. 2000;288(5475):2351-2354

[29] Berglund H et al. The three-dimensional solution structure and dynamic properties of the human FADD death domain. Journal of Molecular Biology. 2000;302(1):171-188

[30] Micheau O, Tschopp J. Induction of TNF receptor I-mediated apoptosis via two sequential signaling complexes. Cell. 2003;114(2):181-190

[31] Eimon PM et al. Delineation of the cell-extrinsic apoptosis pathway in the zebrafish. Cell Death and Differentiation. 2006;13(10):1619-1630

[32] Glenney GW, Wiens GD. Early diversification of the TNF superfamily in teleosts: Genomic characterization and expression analysis. Journal of Immunology. 2007;178(12): 7955-7973

[33] Bobe J, Goetz FW. Molecular cloning and expression of a TNF receptor and two TNF ligands in the fish ovary. Comparative Biochemistry and Physiology. Part B, Biochemistry & Molecular Biology. 2001;129(2-3):475-481

[34] Long Q et al. Stimulation of erythropoiesis by inhibiting a new hematopoietic death receptor in transgenic zebrafish. Nature Cell Biology. 2000;2(8):549-552

[35] Sakata S et al. Conserved function of caspase-8 in apoptosis during bony fish evolution. Gene. 2007;396(1):134-148

[36] Langenau DM et al. Suppression of apoptosis by bcl-2 overexpression in lymphoid cells of transgenic zebrafish. Blood. 2005;105(8):3278-3285

[37] Kwan TT et al. Regulation of primitive hematopoiesis in zebrafish embryos by the death receptor gene. Experimental Hematology. 2006;**34**(1):27-34

[38] Ferrari L et al. FAS/FASL are dysregulated in chordoma and their loss-of-function impairs zebrafish notochord formation. Oncotarget. 2014;**5**(14):5712-5724

[39] Williams JA, Holder N. Cell turnover in neuromasts of zebrafish larvae. Hearing Research. 2000;**143**(1-2):171-181

[40] Viringipurampeer IA et al. Pax2 regulates a fadd-dependent molecular switch that drives tissue fusion during eye development. Human Molecular Genetics. 2012;**21**(10):2357-2369

[41] Henze K, Martin W. Evolutionary biology: Essence of mitochondria. Nature. 2003; **426**(6963):127-128

[42] McBride HM, Neuspiel M, Wasiak S. Mitochondria: More than just a powerhouse. Current Biology. 2006;**16**(14):R551-R560

[43] Brustugun OT et al. Apoptosis induced by microinjection of cytochrome c is caspase-dependent and is inhibited by Bcl-2. Cell Death and Differentiation. 1998;**5**(8):660-668

[44] Kroemer G. Cytochrome c. Current Biology. 1999;**9**(13):R468

[45] Liu X et al. Induction of apoptotic program in cell-free extracts: Requirement for dATP and cytochrome c. Cell. 1996;**86**(1):147-157

[46] Pan Z, Voehringer DW, Meyn RE. Analysis of redox regulation of cytochrome c-induced apoptosis in a cell-free system. Cell Death and Differentiation. 1999;**6**(7):683-688

[47] Zou H et al. Apaf-1, a human protein homologous to C. Elegans CED-4, participates in cytochrome c-dependent activation of caspase-3. Cell. 1997;**90**(3):405-413

[48] Zou H et al. An APAF-1.Cytochrome c multimeric complex is a functional apoptosome that activates procaspase-9. The Journal of Biological Chemistry. 1999;**274**(17):11549-11556

[49] Acehan D et al. Three-dimensional structure of the apoptosome: Implications for assembly, procaspase-9 binding, and activation. Molecular Cell. 2002;**9**(2):423-432

[50] Ikegami R, Hunter P, Yager TD. Developmental activation of the capability to undergo checkpoint-induced apoptosis in the early zebrafish embryo. Developmental Biology. 1999;**209**(2):409-433

[51] Ikegami R et al. Effect of inhibitors of DNA replication on early zebrafish embryos: Evidence for coordinate activation of multiple intrinsic cell-cycle checkpoints at the mid-blastula transition. Zygote. 1997;**5**(2):153-175

[52] Miramar MD et al. NADH oxidase activity of mitochondrial apoptosis-inducing factor. The Journal of Biological Chemistry. 2001;**276**(19):16391-16398

[53] Susin SA et al. Molecular characterization of mitochondrial apoptosis-inducing factor. Nature. 1999;**397**(6718):441-446

[54] Ye H et al. DNA binding is required for the apoptogenic action of apoptosis inducing factor. Nature Structural Biology. 2002;**9**(9):680-684

[55] Li LY, Luo X, Wang X. Endonuclease G is an apoptotic DNase when released from mitochondria. Nature. 2001;**412**(6842):95-99

[56] Du C et al. Smac, a mitochondrial protein that promotes cytochrome c-dependent caspase activation by eliminating IAP inhibition. Cell. 2000;**102**(1):33-42

[57] Verhagen AM et al. Identification of DIABLO, a mammalian protein that promotes apoptosis by binding to and antagonizing IAP proteins. Cell. 2000;**102**(1):43-53

[58] Spiess C, Beil A, Ehrmann M. A temperature-dependent switch from chaperone to protease in a widely conserved heat shock protein. Cell. 1999;**97**(3):339-347

[59] Faccio L et al. Characterization of a novel human serine protease that has extensive homology to bacterial heat shock endoprotease HtrA and is regulated by kidney ischemia. The Journal of Biological Chemistry. 2000;**275**(4):2581-2588

[60] Gray CW et al. Characterization of human HtrA2, a novel serine protease involved in the mammalian cellular stress response. European Journal of Biochemistry. 2000; **267**(18):5699-5710

[61] Suzuki Y et al. A serine protease, HtrA2, is released from the mitochondria and interacts with XIAP, inducing cell death. Molecular Cell. 2001;**8**(3):613-621

[62] Tsujimoto Y et al. Involvement of the bcl-2 gene in human follicular lymphoma. Science. 1985;**228**(4706):1440-1443

[63] Tsujimoto Y, Croce CM. Molecular cloning of a human immunoglobulin lambda chain variable sequence. Nucleic Acids Research. 1984;**12**(22):8407-8414

[64] Tsujimoto Y et al. The t(14;18) chromosome translocations involved in B-cell neoplasms result from mistakes in VDJ joining. Science. 1985;**229**(4720):1390-1393

[65] Cleary ML, Smith SD, Sklar J. Cloning and structural analysis of cDNAs for bcl-2 and a hybrid bcl-2/immunoglobulin transcript resulting from the t(14;18) translocation. Cell. 1986;**47**(1):19-28

[66] Youle RJ, Strasser A. The BCL-2 protein family: Opposing activities that mediate cell death. Nature Reviews. Molecular Cell Biology. 2008;**9**(1):47-59

[67] Kratz E et al. Functional characterization of the Bcl-2 gene family in the zebrafish. Cell Death and Differentiation. 2006;**13**(10):1631-1640

[68] Inohara N, Nunez G. Genes with homology to mammalian apoptosis regulators identified in zebrafish. Cell Death and Differentiation. 2000;**7**(5):509-510

[69] Chen MC et al. Cloning and characterization of a novel nuclear Bcl-2 family protein, zfMcl-1a, in zebrafish embryo. Biochemical and Biophysical Research Communications. 2000;**279**(2):725-731

[70] Prudent J et al. Bcl-wav and the mitochondrial calcium uniporter drive gastrula morphogenesis in zebrafish. Nature Communications. 2013;**4**:2330

[71] Zhong JX et al. Zebrafish Noxa promotes mitosis in early embryonic development and regulates apoptosis in subsequent embryogenesis. Cell Death and Differentiation. 2014; **21**(6):1013-1024

[72] Arnaud E et al. The zebrafish bcl-2 homologue Nrz controls development during somitogenesis and gastrulation via apoptosis-dependent and -independent mechanisms. Cell Death and Differentiation. 2006;**13**(7):1128-1137

[73] Popgeorgiev N et al. The apoptotic regulator Nrz controls cytoskeletal dynamics via the regulation of Ca^{2+} trafficking in the zebrafish blastula. Developmental Cell. 2011;**20**(5):663-676

[74] Bonneau B et al. The Bcl-2 homolog Nrz inhibits binding of IP3 to its receptor to control calcium signaling during zebrafish epiboly. Science Signaling. 2014;**7**(312):ra14

[75] Prudent J, Gillet G, Popgeorgiev N. Nrz but not zBcl-xL antagonizes Bcl-wav pro-apoptotic activity in zebrafish. Communicative & Integrative Biology. 2014;**7**(1):e28008

[76] Prudent J et al. Bcl-2 proteins, cell migration and embryonic development: lessons from zebrafish. Cell Death & Disease. 2015;**6**:e1910

[77] Cheng R et al. Zebrafish (*Danio rerio*) p53 tumor suppressor gene: cDNA sequence and expression during embryogenesis. Molecular Marine Biology and Biotechnology. 1997; **6**(2):88-97

[78] Thisse C et al. The Mdm2 gene of zebrafish (Danio Rerio): Preferential expression during development of neural and muscular tissues, and absence of tumor formation after over-expression of its cDNA during early embryogenesis. Differentiation. 2000;**66**(2-3):61-70

[79] Lee KC et al. Detection of the p53 response in zebrafish embryos using new monoclonal antibodies. Oncogene. 2008;**27**(5):629-640

[80] Berghmans S et al. tp53 mutant zebrafish develop malignant peripheral nerve sheath tumors. Proceedings of the National Academy of Sciences of the United States of America. 2005;**102**(2):407-412

[81] den Hertog J. Tumor suppressors in Zebrafish: From TP53 to PTEN and beyond. Advances in Experimental Medicine and Biology. 2016;**916**:87-101

[82] Storer NY, Zon LI. Zebrafish models of p53 functions. Cold Spring Harbor Perspectives in Biology. 2010;**2**(8):a001123

[83] Bladen CL et al. DNA damage response and Ku80 function in the vertebrate embryo. Nucleic Acids Research. 2005;**33**(9):3002-3010

[84] Liu TX et al. Knockdown of zebrafish Fancd2 causes developmental abnormalities via p53-dependent apoptosis. Developmental Cell. 2003;**5**(6):903-914

[85] Fischer S et al. Mutation of zebrafish caf-1b results in S phase arrest, defective differentiation, and p53-mediated apoptosis during organogenesis. Cell Cycle. 2007;**6**(23):2962-2969

[86] Ekker SC, Larson JD. Morphant technology in model developmental systems. Genesis. 2001;**30**(3):89-93

[87] Robu ME et al. p53 activation by knockdown technologies. PLoS Genetics. 2007;**3**(5):e78

[88] Bill BR et al. A primer for morpholino use in zebrafish. Zebrafish. 2009;**6**(1):69-77

[89] Hensey C, Gautier J. A developmental timer that regulates apoptosis at the onset of gastrulation. Mechanisms of Development. 1997;**69**(1-2):183-195

[90] Alli Shaik A et al. Functional mapping of the zebrafish early embryo proteome and transcriptome. Journal of Proteome Research. 2014;**13**(12):5536-5550

[91] Popgeorgiev N et al. The yolk cell of the zebrafish blastula harbors functional apoptosis machinery. Communicative & Integrative Biology. 2011;**4**(5):549-551

[92] Kimmel CB, Law RD. Cell lineage of zebrafish blastomeres. II. Formation of the yolk syncytial layer. Developmental Biology. 1985;**108**(1):86-93

6

The Role of PSR in Zebrafish (*Danio rerio*) at Early Embryonic Development

Wan-Lun Taung, Jen-Leih Wu and Jiann-Ruey Hong

Abstract

During development, the role of the phosphatidylserine receptor (PSR) in the professional removal of apoptotic cells that have died is few understood. Programmed cell death (PCD) began during the shield stage (5.4 hpf), with dead cells being engulfed by a neighboring cell that showed a normal-looking nucleus and the nuclear condensation multi-micronuclei of an apoptotic cell. Recently, in the zebrafish model system, PS receptor played a new role on corpse cellular cleaning for further normal development during early embryonic development, which also correlated with tissues' or organs' complete development and organogenesis. In the present, we summary new story that a transcriptional factor, YY1a, in the upstream of PSR is how to regulate PS receptor expression that linked to function of PSR-phagocyte mediated apoptotic cell engulfment during development, especially the development of organs such as the brain and heart. YY1a/PSR-mediated engulfing system may involve in diseases and therapy. This engulfing system may provide new insight into phosphatidylserine receptor how to dynamitic interaction with apoptotic cell during priming programmed cell death.

Keywords: programmed cell death, apoptosis, phosphatidylserine receptor, early embryonic development, brain, *in vivo* rescued

1. Introduction

Apoptotic cell death occurs by a mechanism that is conserved from nematodes to humans [1]. *In vivo*, the typical fate for apoptotic cells is rapid engulfment and degradation by phagocytes [2]. Among higher organisms, the removal of apoptotic cells by phagocytes suppresses inflammation, modulates the macrophage-directed deletion of host cells, and critically regulates the immune response of an individual [3]. Cell death that is morphologically and genetically distinct from apoptosis is strongly implicated in some human diseases [1].

For vertebrates, the phagocyte engages the dying cells through specific receptors that include the phosphatidylserine receptor (PSR) [4–6], complement receptors 3 and 4, the ABC1 transporter, and members of the scavenger-receptor family [7]. Recently, T-cell immunoglobulin mucin protein 4 (TIM4), a phosphatidylserine (PtdSer)-binding receptor, mediates the phagocytosis of apoptotic cells. TIM4 engages integrins as co-receptors to evoke the signal transduction needed to internalize PtdSer-bearing targets such as apoptotic cells [8]. And PSR-1 enriches and clusters around apoptotic cells during apoptosis. These results establish that PSR-1 is a conserved, phosphatidylserine-recognizing phagocyte receptor [9].

For non-vertebrate systems such as the nematode Caenorhabditis elegans [10, 11] and *Drosophila melanogaster* [12], it illustrates the power of using genetically tractable systems to identify necessary phagocytic genes. Major efforts to understand crucial pathways that mediate programmed cell death (PCD) have also led to the genetic and molecular characterization of a number of genes involved in the recognition and engulfment mechanisms of cells among invertebrates [10, 13–15]. For *Caenorhabditis elegans*, it is important to recognize that phagocytosis is performed by cells that are non-specific phagocytes rather than by specialized phagocytes such as macrophages, as tends to be the case in *Drosophila melanogaster* [3].

For lower vertebrate systems such as the zebrafish, the cell corpses generated developmentally are quickly removed, although which specific type(s) of engulfment genes are involved still remain largely unknown. Little is known regarding the molecular mechanisms by which the resulting (cell) corpses are eliminated and the clearance of defective events for zebrafish. The zebrafish PSR-engulfing receptor was cloned (zfpsr) by Hong et al. [16], and its nucleotide sequence, was compared with corresponding sequences in Drosophila melanogaster (76% comprising identity), human (74%), mouse (72%), and *Caenorhabditis elegans* (60%). The PSR receptor contained a JmjC domain (residues 143–206), and localization was labeled in chromosome 3 (GCF-000002035.6; accession number: NC-007114.7). Very recently, the PSR gene was regulated by YY1a transfection factor [17].

2. What is programmed cell death?

The concept of natural cell death can go back to 1842 [18]. Karl Vogt found that mid-wife toad eliminates notochord and forms vertebrae during metamorphosis. The death of this cell depends on the regulation of endogenous genes and hence giving rise to the term programmed cell death (PCD). Now the PCD has generated a new concept and new clarification.

2.1. Type I cell death: apoptosis

Apoptosis, a type of programmed cell death, is a mechanism in developing embryos that removes damaged cells without impairing the overall development of normal tissues [19]. Controlling factors of apoptosis include the B-cell lymphoma 2 (Bcl-2) protein family that inhibits apoptosis, the Bcl-2-associated X protein (Bax) protein that promotes apoptosis, and the aspartate specific cysteine protease (Caspase) family of proteins [20, 21]. Regulation of apoptosis includes regulation both inside and outside. Among them, the activation of the external is mainly when the death ligand on the cell membrane binds with the death receptor, and the apoptosis pathway is activated, which in turn activates the downstream Caspase-8 and the downstream Caspase-3 to promote apoptosis. Internal activation is mediated mainly by endogenous stimuli such as DNA damage, Bax, and Bcl-2 homologous antagonist killer protein (Bak) that cause the activation of pro-apoptotic Bcl-2 family in the mitochondrial outer membrane, resulting in grain line and then release of cytochrome C to combine with apoptotic protease activating factor 1 (Apaf-1) to form apoptotic body (apoptosome), which in turn activates downstream Caspase-9 and downstream Caspase-3 to promote cell apoptosis [21].

2.2. Type II cell death: autophagy

Autophagy is a catabolic process that involves the degradation of cytoplasmic components, protein aggregates, and organelles through the formation of autophagosomes, which are degraded by fusion with lysosomes. The autophagy process has been extensively well studied in the response to starvation of *Saccharomyces cerevisiae*, in which it protects cells from death by recycling cell contents. Autophagy depends on a large group of evolutionarily conserved autophagy-related genes (ATG) [22]. On the other hand, the protective, pro-survival function of autophagy, silencing, and deletion of ATG genes that resulted in accelerated cell death [23, 24] was proposed. However, in certain scenarios, it has been suggested that severely triggered autophagy process can lead to or contribute to cell death.

2.3. Type III cell death: regulated necrosis

In the early stage, necrosis was regarded as an unregulated mode of cell death that was caused by overwhelming trauma. However, many recent studies indicate the existence of several modes of regulated necrosis [25]. Necrosis is characterized by swelling of organelles and cells, rupture of the plasma membrane, and release of the intracellular contents. Different modes of regulated necrosis share common morphological features. Then, the best well-studied form of regulated necrosis, also called necroptosis, is a type of necrotic cell death that depends on receptor-interacting serine/threonine-protein kinase 1 (RIPK1) and/or RIPK3 [25–27]. Additional necrosis is induced by different stimuli, but it remains to be shown that these actually involve different mechanisms for programmed necrosis [25].

3. The zebrafish development in the early stage

Zebrafish embryos develop from the one-cell stage after fertilization. The cells are then split in multiples to blastocysts' stage (see **Figure 1**); then enters the gastrula stage. During this period, embryonic cells begin to differentiate into three germ layers via apoptosis. Each germ layer will differentiate into specific organs such as endoderm cells that differentiate into respiratory and gastrointestinal epithelial linings and include glandular cells such as

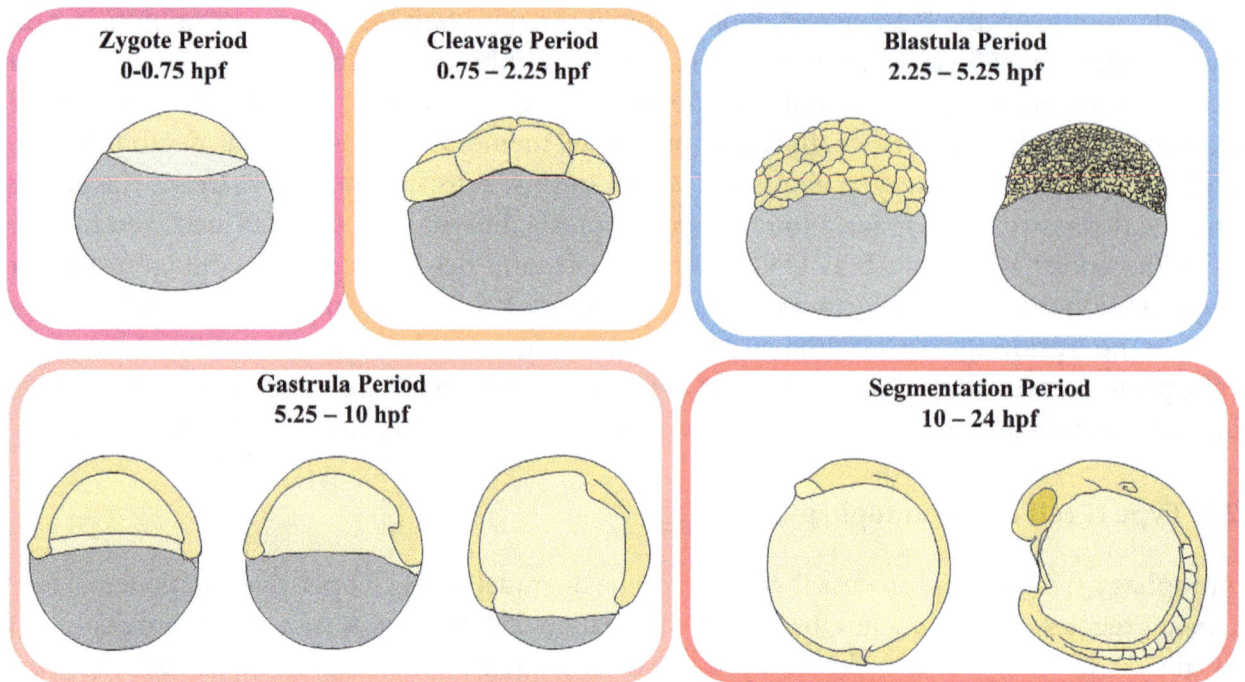

Figure 1. A scheme of zebrafish embryonic development stages was shown. After fertilization, one-cell formation in animal pole is for about 30 min and then begins dividing into two cells (0.5–1.0 hpf) and entering into blastula stage (2.25–5.25 hpf), gastrula stage (5.25–10 hpf), and segmentation stage (10–24 hpf) that finally can hatch out between 48 and 72 hpf. In the right time in the gastrula stage at 5.4–6.0 hpf, the programmed cell death is turned on and the professional engulfing corpses system by PSR works for smoothing embryonic development.

the liver and pancreas of the relevant organs. The mesoderm will differentiate into smooth muscle layers, smooth muscle coats, connective tissues, and blood vessels that supply these organs. It is also a source of blood cells, bone marrow, bone, striated muscles, reproductive, and excretory organs. This period of cell differentiation behavior and follow-up organ development are closely related. Then enter the segmentation, pharynx, and incubation periods (**Figure 1**) [16].

4. The role of apoptosis in early developmental stage in zebrafish

In the zebrafish system, apoptosis is regulated by Noxa, which is a novel regulator of early mitosis before the 75% epiboly stage when it translates into a key mediator of apoptosis in subsequent embryogenesis [28]. PCD turned on was observed by electron microscopy and the earliest onset of programmed cell death in zebrafish embryos was about 6 h after fertilization, and typical apoptosis was observed in zebrafish embryos in the gastrulation stage [16]. In addition, 12–96 h after fertilization, signals of apoptotic cells can be detected in the nervous system and sensory organs such as the retina, ear, and olfactory organs [16].

5. After PCD starting: engulfing of copses death cell by professional system, PSR and others

The *Caenorhabditis elegans* system is known to have a nonspecific phagocytic system that regulates cell death through the use of cell death abnormality protein 1 (CED-1) to identify cellular debris and transmit related messages and another group of cell death. The system of abnormality protein 2 (CED-2) affects the cytoskeleton, causing the cell membrane to collapse, while activating the relevant GTP synthetase to generate sufficient energy to provide cellular pattern changes. By this system, Wang et al. confirmed that PSR-1 of the nematode activates CED-2, further enabling phagocytes to recognize apoptotic bodies and using enzyme immunoassay to prove that PSR-1 itself can interact with phosphatidylserine [11].

Proliferation of programmed cell death (PCD) is caused by the condensation of chromatin DNA, which leads to the cleavage of DNA in the cell and the cleavage of the membrane by the nuclear pore to form nuclear fragments. In the process of programmed death, dehydration will continue, the cytoplasm is concentrated, resulting in vesicle-like cell membrane, and cell size decreases. Apoptotic cells produce nuclear fragmentation and form chromatin masses (nuclear fragments), resulting in sprouting of cells, the formation of a spherical bulge, and other means of cell protrusions, eventually resulting in a range of sizes, including the cytoplasm, organelles and nuclear debris, and other small bodies, which can be called apoptotic bodies [29]. In general, there are three main components of the phospholipids in the cell membrane, including phosphatidylcholine (PC), phosphatidylethanolamine (PE), and phosphatidylserine (PS) located on the inner side of the cell membrane [30]. When cells undergo apoptosis, the phospholipid structure inside the cell membrane moves to the outside of the cell membrane, at which time the PS located on the surface of the cell membrane becomes an important marker of the phagocytic apoptotic bodies [31, 32]. Scientists with competing PS analogs competed for the ability to inhibit PS clearance of apoptotic cells, but failed to do so for other phospholipids, such as PC, confirming that phospholipid serine (PS) can be specifically affected by identification [32]. Fadok et al. further induced macrophages to produce PSR antibody mAb 217 with transforming growth factor beta (TGF-beta) and beta-glucan [4]. mAb 217 can be used to calibrate phagocytes with the ability to recognize, and vice versa phagocytes without identifying apoptotic bodies, so that the monoclonal antibody can identify possible target proteins and purify the deglycosylation to obtain the target protein of 48 kDa, PSR [4]. However, previous studies on phospholipid serine receptors (PSRs) have shown a nuclear localization of phospholipid serine receptors mainly in vertebrate cell lines and invertebrates [33]. Then, by comparing the results of amino acid sequence-predicting phospholipids, serine receptors may have DNA binding structures and Jumonji C domain (JmjC domain); the results of this comparison in different species, including from hydra to humans, all have a certain conservative [33, 34]. In addition, nematodes [4, 11], fruit flies [4], zebrafish [16], and mouse [6] also exist in the PS receptor.

6. The role of zebrafish PSR in early embryonic development

In the zebrafish system, Hong et al. [16] compared the PSR genes of different species in the cDNA library to further identify the zfpsr homology gene of PSR in zebrafish. After the zebrafish embryos were fertilized, by in situ hybridization, to observe the zfpsr development in whole embryos, zfpsr mRNA was found in one-cell embryo and then displayed in different tissues and organs as time progressed [16]. At 24 h (see **Figure 2**) after fertilization, the head, eyes, body axis, and chordal can be observed in the performance of the gene. After 3 days of fertilization, zfpsr mRNA can be observed in specific organs such as the heart, trunk, kidney, and other organs in the development of the gastrointestinal tract and is the predominant organ of zfpsr. Then, the zfpsr gene is silenced (loss of function by a morpholino), resulting in the accumulation of apoptotic cells in zebrafish development, further resulting in embryonic brain, heart, chordate, somite dysplasia. In severely deficient embryos, the brain is impaired, there incomplete development of the posterior nodules, and an inability to hatch. Slightly deficient embryos developed in the heart and in the absence of the apical development, including the heart chambers and the aorta. Large veins underwent incomplete development. The severely affected embryos accumulated large amounts of cellular debris 12 h after fertilization and died 3 days after fertilization.

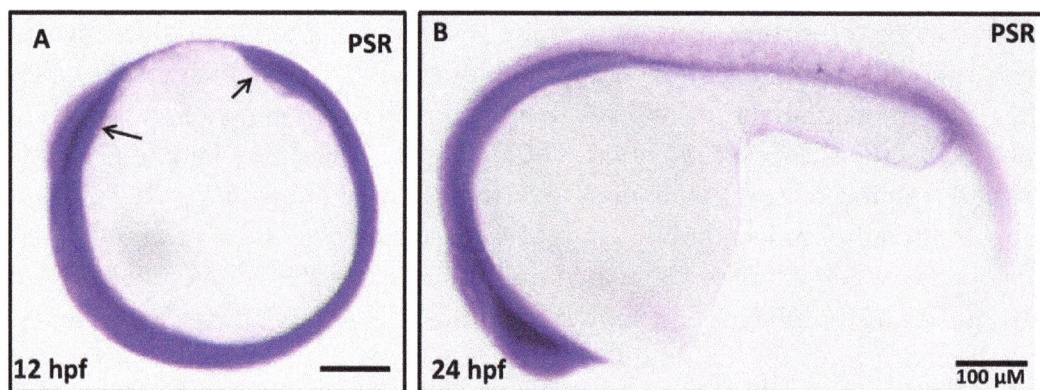

Figure 2. Identification of PSR expression pattern during early zebrafish embryonic development. (A) PSRs are expressed in the whole embryo including the ectoderm, mesoderm, and endoderm, especially with the major location for PSR being within the brain region and the posterior of the embryo (indicated by arrows) at 12 hpf and (B) at 24 hpf, PSR is expressed in the whole notochord and distributed in the trunk, brain, and the eyes. Scale bar = 100 μm.

7. Why PSR is important in zebrafish?

Previous studies indicated that the environmental stress in zebrafish embryos may be related to the regulation of mitogen-activated protein kinase. And mitogen-activated protein kinases have been shown to regulate cellular migration in previous studies [35, 36], suggesting that adversity within the embryo may affect cell migration. In our laboratory, it was observed that the zebrafish embryos after PSR knockout were observed in the early stage of embryonic development of the intestine, and the laryngeal phenomenon of outsourcing was observed

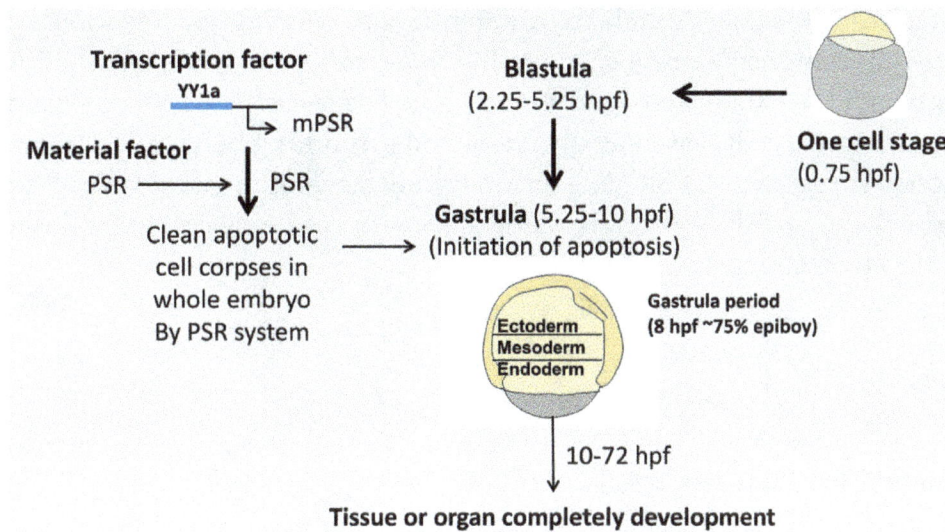

Figure 3. A summary of the zebrafish PSR story during the entire development. The role of PSR function, especially in the gastrula stage of zebrafish embryo, is very important to the event of epiboly. The programmed cell death star via apoptosis may be at 75% epiboly stage as compared to PSR knockdown control. During this stage, the cell death should be cleared out by the engulfing system that may be working through the PSR engulfing system. If not, dead cells will accumulate and cause normal cell migration. At the same time, the accumulated corpus cell can enhance environmental stress via reactive oxygen species (ROS) production in the whole embryo. Finally, these reasons may cause the phenotype of cell fate disruption within three germ layers and continue till late development stage.

(see **Figure 3**), and the differentiation of the three germ layers was not clear. Previous studies have also confirmed that the PSR gene may be involved in the clearance of dead cells [16]. Our study delayed the migration of PSR cells after they were deactivated, which might be due to the unclear death cells, which blocked the movement of the whole cell layer and delayed the development of zebrafish embryos in the primitive intestine. However, the mechanism is still not clear. By virtue of the early embryonic development of endoderm-labeled embryonic forerunner cells, progenitor cells are severely damaged after PSR loss of function and lead to disruption of the differentiation of the cell cycle to cell migration. It is concluded that the PSR gene affects the migration and differentiation of cells in the embryonic development of the intestine, leading to the inability of cells to determine the fate of the cells in the early development of the embryo and disrupting the distribution of the germinal layers. For new supporting case, such as skeletal muscle arises from the fusion of precursor myoblasts into multinucleated myofibers. A new report by Hochreiter-Hufford et al. [37] identifies apoptotic cells as a new type of cue that induces signaling via the phosphatidylserine receptor BAI1 to promote fusion of healthy myoblasts, with important implications for muscle development and repair.

8. Conclusion and perspectives

In gastrula stages of zebrafish early embryo development, germ layer differentiation is quite important. This chapter is about the formation of follow-up organs. Differentiation of the germ layer requires apoptosis to assist in participation. According to the results of our laboratory research, the loss of function of the PSR gene by knockout approach can result in the failure of early gastrula-stage germ layer differentiation, which in turn led to the increase of oxidative stress in zebrafish embryos, triggering severe apoptosis (unpublished data). Then, PSR knockout induced some damaged organs and tissues which was observed at 72 hpf that was also linked to cardiovascular dysplasia and swimming behavior. A summary of the abovementioned suggests that PSR gene on embryonic development and organ development has a certain impact, which is strongly associated with cardiovascular dysplasia and brain development even on congenital diseases.

Author details

Wan-Lun Taung[1,2], Jen-Leih Wu[3] and Jiann-Ruey Hong[1,2*]

*Address all correspondence to: jrhong@mail.ncku.edu.tw

1 Laboratory of Molecular Virology and Biotechnology, Institute of Biotechnology, National Cheng Kung University, Tainan City, Taiwan (R.O.C)

2 Department of Biotechnology and Bioindustry, National Cheng Kung University, Tainan City, Taiwan (R.O.C)

3 Laboratory of Marine Molecular Biology and Biotechnology, Institute of Cellular and Organismic Biology, Taipei, Taiwan (R.O.C)

References

[1] Meier P, Finch A, Evan G. Apoptosis in development. Nature. 2000;**407**:796-801

[2] Savill J. Apoptosis: Phagocytic docking without shocking. Nature. 1998;**392**:442-443

[3] Savill J, Fadok V. Corpse clearance of defines the meaning of cell death. Nature. 2000; **407**:784-788

[4] Fadok VA, Bratton DL, Rose DM, Pearson A, Ezekewitz RA, Henson PM. A receptor for phosphatidylserine-specific clearance of apoptotic cells. Nature. 2000;**405**:85-90

[5] Hong JR, Lin TL, Hsu YL, Wu JL. Apoptosis procedes necrosis of fish cell line by infectious pancreatic necrosis virus. Virology. 1998;**250**:76-84

[6] Li MO, Sarkisian MR, Mehal WZ, Rakic P, Flavell RA. Phosphatidylserine receptor is required for clearance of apoptotic cells. Science. 2003;**302**:1560-1563

[7] Platt N, da Silva RP, Gordon S. Recognizing death: The phagocytosis of apoptotic cells. Trends in Cell Biology. 1998;**8**:365-372

[8] Flannagan RS, Canton J, Furuya W, Glogauer M, Grinstein S. The hosphatidylserine receptor TIM4 utilizes integrins as coreceptors to effect phagocytosis. Molecular Biology of the Cell. 2014;**25**(9):1511-1522

[9] Yang H, Chen YZ, Zhang Y, Wang X, Zhao X, et al. A lysine-rich motif in the phosphatidylserine receptor PSR-1 mediates recognition and removal of apoptotic cells. Nature Communications. 2015;7(6):5717

[10] Chung S, Gumienny TL, Haengartner MO, Driscoll M. A common set of engulfment genes medistes removal of both apoptotic and necrotic cell corpses in *C. elegans*. Nature Cell Biology. 2000;2:931-937

[11] Wang X, Wu YC, Fadok VA, Lee MC, Keiko GA, Cheng LC, Ledwich D, Hsu PK, Chen JY, Chou BK, et al. Cell corpse engulfment mediated by *C. elegans* phosphatidylserine receptor through CED-5 and CED-12. Science. 2003;302:1563-1566

[12] Franc NC, Heitzler P, Ezekowitz AB, White K. Requirement for croquemort in phagocytosis of apoptotic cells in drosophila. Science. 1999a;284:1991-1994

[13] Lauber K, Bohn E, Krober SM, Xiao YJ, Blumenthal SG, Lindemann RK, Marini P, Wiedig C, Zobywalski A, Baksh S, et al. Apoptotic cells induce migration of phagocytes via caspase-3-mediated release of a lipid attraction signal. Cell. 2003;113:717-730

[14] Arur S, Uche UE, Rezaui MF, Scranton V, Cowan AE, Mohler W, Han DK. Annexin I is an endogenous ligand that mediates apoptotic cell engulfment. Developmental Cell. 2003;4:587-598

[15] Ravichandra KS. Recruitment signals from apoptotic cells: Invitation to a quiet meal. Cell. 2003;113:817-820

[16] Hong JR, Lin GH, Lin CJ, Wang WP, Lee CC, Lin TL, Wu JL. Phosphatidylserine receptor is required for the engulfment of dead apoptotic cells and for normal embryonic development in zebrafish. Development. 2004;131:5417-5427

[17] Shiu WL, Huang KR, Hung JC, Wu JL, Hong JR. Knockdown of zebrafish YY1a can downregulate the phosphatidylserine (PS) receptor expression, leading to induce the abnormal brain and heart development. Journal of Biomedical Science. 2016;23:31

[18] Vogt CI. Untersuchungen über die Entwicklungsgeschichte der Geburtshelferkröte (Alytes obstetricans) (in German) (Jent, 1842)

[19] Jacobson MD, Weil M, Raff MC. Programmed cell death in animal development. Cell. 1997;88:347-354

[20] Youle RJ, Strasser A. The BCL-2 protein family: Opposing activities that mediate cell death. Nature Reviews. Molecular Cell Biology. 2008;9:47-59

[21] Fuchs Y, Steller H. Live to die another way: Modes of programmed cell death and the signals emanating from dying cells. Nature Reviews. Molecular Cell Biology. 2015; 16:329-244

[22] Mizushima N, Komatsu M. Autophagy: Renovation of cells and tissues. Cell. 2011; 147:728-741

[23] Maiuri MC, Zalckvar E, Kimchi A, Kroemer G. Self-eating and self-killing: Crosstalk between autophagy and apoptosis. Nature Reviews. Molecular Cell Biology. 2007;8:741-752

[24] Levine B, Yuan J. Autophagy in cell death: An innocent convict? The Journal of Clinical Investigation. 2005;**115**:2679-2688

[25] Van den Berghe T, Linkermann A, Jouan- Lanhouet S, Walczak H, Vandenabeele P. Regulated necrosis: The expanding network of non-apoptotic cell death pathways. Nature Reviews. Molecular Cell Biology. 2014;**15**:135-147

[26] Galluzzi L, Kroemer G. Necroptosis: A specialized pathway of programmed necrosis. Cell. 2008;**135**:1161-1163

[27] Vandenabeele P, Galluzzi L, Vanden Berghe T, Kroemer G. Molecular mechanisms of necroptosis: An ordered cellular explosion. Nature Reviews. Molecular Cell Biology. 2010;**11**:700-714

[28] Zhong JX, Zhou L, Li Z, Wang Y, Gui JF. Zebrafish Noxa promotes mitosis in early embryonic development and regulates apoptosis in subsequent embryogenesis. Cell Death and Differentiation. 2014;**21**(6):1013-1024

[29] Lawen A. Apoptosis-an introduction. BioEssays. 2003;**25**:888-896

[30] Bretscher MS. Asymmetrical lipid bilayer structure for biological membranes. Nature: New Biology. 1972;**236**:11-12

[31] Voll RE, Herrmann M, Roth EA, Stach C, Kalden JR, Girkontaite I. Immunosuppressive effects of apoptotic cells. Nature. 1997;**390**:350-351

[32] Fadok VA, Bratton DL, Konowal A, Freed PW, Westcott JY, Henson PM. Macrophages that have ingested apoptotic cells in vitro inhibit proinflammatory cytokine production through autocrine/paracrine mechanisms involving TGF-beta, PGE2, and PAF. Journal of Clinical Investigation. 1998;**101**:890-898

[33] Cikala M, Alexandrova O, David CN, Pröschel M, Stiening B, Cramer P, Böttger A. The phosphatidylserine receptor from hydra is a nuclear protein with potential Fe (II) dependent oxygenase activity. Biomed central genomics cell. Biology. 2004;**5**:26

[34] Clissold PM, Ponting CP. JmjC: Cupin metalloenzyme-like domains in jumonji, hairless and phospholipase a 2 β. Trends in Biochemical Sciences. 2001;**26**(1):7-9

[35] Shi X, Zhou B. The role of Nrf2 and MAPK pathways in PFOS-induced oxidative stress in zebrafish embryos. Toxicological Sciences. 2010;**115**(2):391-400

[36] Tseng HL, Li CJ, Huang LH, Chen CY, Tsai CH, Lin CN, Hsu HY. Quercetin 3-O-methyl ether protects FL83B cells from copper induce oxidative stress through the PI3K/Akt and MAPK/Erk pathway. Toxicology and Applied Pharmacology. 2012;**264**(1):104-113

[37] Hochreiter-Hufford AE, Lee CS, Kinchen JM, Sokolowski JD, Arandjelovic S, Call JA, Klibanov AL, Yan Z, Mandell JW, Ravichandran KS. Phosphatidylserine receptor BAI1 and apoptotic cells as new promoters of myoblast fusion. Nature. 2013;**497**(7448):263-267

Zebrafish Aging Models and Possible Interventions

Dilan Celebi-Birand, Begun Erbaba,
Ahmet Tugrul Ozdemir, Hulusi Kafaligonul and
Michelle Adams

Abstract

Across the world, the aging population is expanding due to an increasing average life expectancy. The percentage of elderly over the age of 65 is expected to be more than 15% of the total world population by 2025. As the lifespan increases, there will be a need for maintaining a healthy state for these individuals. Our current knowledge on types and durations of potential anti-aging therapies is quite limited. Recently the zebrafish has emerged as a promising model for understanding the cognitive and neurobiological changes during aging, as well as its use with potential anti-aging interventions. Like humans this model organism ages gradually, displays similar behavioral properties and social characteristics, and in addition, there is a wealth of molecular and genetic tools to uncover the cellular mechanism that contribute to age-related cognitive declines. Drug effect and toxicity can be easily tested in the zebrafish. Therefore, this animal model can provide information about potential therapies that could be translated directly into human populations or provide a more focused treatment direction for testing in other mammalian animal models. The zebrafish will be a powerful tool for uncovering the mysteries of the aging brain.

Keywords: aging, behavior, neurobiology, dietary regimens, rapamycin, morpholino, zebrafish aging models

1. Introduction: the zebrafish as a model organism for brain aging

Throughout history, humans have tried to find ways to delay or reverse aging and cure age-related diseases. Whether it was a dream of finding the "fountain of youth" or use of modern applications such as drugs with alleged "anti-aging" properties; to date, there are no interventions that met the expectations of the humankind. Therefore, we should first understand the complex multifactorial nature of aging and its underlying mechanisms before we aim to intervene. Both genetic and epigenetic factors are involved in aging [1, 2] and this holds true across different species.

Extensive investigations have focused on the mechanisms of aging by using both vertebrate and invertebrate animal models to provide an insight into age-related physiological, cognitive and neurobiological changes, with each giving a piece of the puzzle. Some of the animal models that have been studied in the context of aging are worms (e.g., *Caenorhabditis elegans*) [3], fruit flies (e.g., *Drosophila melanogaster*) [4], mice (*Mus musculus*) [5], and non-human primates (e.g., rhesus monkeys) [6–9]. Rhesus monkeys provided valuable information about the mechanisms that underlie physiological and cognitive changes related to age [9], whereas genetic screens on invertebrate models yielded a list of genes involved in the regulation of life span [6, 7]. All these animal models have played important roles in understanding mammalian aging and age-related diseases but having only one animal model might be limiting. For example, invertebrate models have a short lifespan, which might not be correctly translated to humans, and genes that are associated with longevity in vertebrates with longer lifespans might not be revealed by studies on these models. Non-human primates, on the other hand, have a lifespan of about 25 years, making longitudinal studies very difficult. Mice are nocturnal animals while humans are diurnal and their diverse circadian rhythms may affect the aging process differentially, which might complicate the translation of the knowledge obtained from mice studies into humans. Thus, recently the zebrafish (*Danio rerio*) has emerged as a novel model for vertebrate aging research [10–15]. While it has been used in numerous fields previously due to its optical transparency during early development, cost-effectiveness, high fecundity, detailed characterization of its genome, diverse mutant or transgenic strains that have been made available; it was the emergence of evidence for the gradual aging phenotype in zebrafish and its diurnal nature that drew attention from researchers studying normal aging.

The zebrafish has on average a maximum lifespan of 3 to 5 years in laboratory conditions. The gradual senescence phenotype allowed researchers to identify age-related changes in gene expression [16], endocrine [14] and neuroendocrine system [17], musculoskeletal [10], visual function and morphology [18], cognitive functions [14], and sleep [13, 19]. In addition to studies monitoring changes in zebrafish from early development to old age [13, 19, 20], zebrafish has become a model for assessment of behavior and cognitive functions. There are several cognitive tests available for use on zebrafish to index learning and memory [21–23], anxiety and stress response [24]. The use of zebrafish in cognitive and neurobiological aging research increased with the accumulation of knowledge about the neuroanatomy of the zebrafish brain. The standard neuroanatomical, neuropharmacological and immunohistochemical techniques have been applied to characterize the organization of the zebrafish brain [25, 26]. These studies demonstrated similarities of zebrafish sensory and motor systems, and central nervous system circuits with other vertebrates. In addition to homology at systemic level, homologous structures such as zebrafish lateral pallium and mammalian hippocampus have been shown based on electrophysiological [27] and neurochemical data [28]. Lateral pallium is particularly important due to its suggested role in learning and memory in zebrafish [29], processes that are deeply affected by age. While understanding the age-related physiological and cognitive changes remain crucial, studies involving interventions that might delay or reverse these changes run parallel. Zebrafish is an outstanding model for investigating drug toxicity and/or effect throughout lifespan, identification of age-associated biomarkers that could become predictors of premature aging phenotypes, and screening mutants for identification of strains with accelerated or decelerated aging phenotypes. In this chapter, we will review age-related behavioral and neurobiological changes in zebrafish and continue with existing models that have delayed or accelerated aging phenotype.

2. Age-related changes in behavior and biology

2.1. Behavioral changes

2.1.1. Age-related changes in human perceptual and cognitive performance

Vision is the most informative of our senses and hence, essential for survival in a dynamic world. By relying on vision, we are able to recognize visual objects in the environment, judge the trajectories of fast approaching objects and cruise through morning traffic. Therefore, most of the aging studies on human perception focused on visual perception and cognition. A number of studies have shown that visual functioning is significantly altered throughout aging. Older adults have typically impaired visual sensitivity and altered perception of different visual features such as motion [30]. As opposed to the traditional view focusing on structural changes in the eye and retina, accumulating evidence suggests that impairments in neural circuitry and functioning in the cortex have important contributions to the age-related changes in visual sensitivity and perception [31]. For instance, behavioral studies have shown that older adults are less accurate in discriminating motion direction and speed compared to younger adults [32, 33]. Moreover, older adults typically need more time to make visual judgments. This suggests that older population have also slowed visual processing speed. In line with these changes in perceptual performance, neurophysiological studies on different species have shown that neurons located in visual area V1 and MT have less direction and speed sensitivity due to aging [34, 35]. These changes in the cortical neurons have been mostly explained by the deterioration in synaptic connections and integrity, and hence increase in the noise level of the local cortical network.

Normal aging is most notably accompanied by declines in cognitive functions and processes. Accumulating evidence suggests that age-related decline exists in both low-level and high-level cognitive processes [36]. Previous studies have pointed out a general deceleration in cognitive processes and a decline in attentional resources due to aging [37, 38]. Moreover, there exist age-correlated deficits in learning, memory and cognitive control. Age-related declines in cognitive function are typically reflected as significant losses of learning and memory abilities [39]. Though decrements in implicit and short-term memory tasks are mostly slight, age-related performance declines in working, episodic and prospective memory tasks are substantial [40]. In older adults, the reduction in memory context and false recollections are also commonly found. Another common observation is that older adults have difficulty in associating different aspects of an event. It should be also noted that some cognitive functions can remain intact and may even improve (e.g., semantic knowledge) throughout aging. It was initially thought that age-related cognitive decline was due to massive loss of neurons. However, current research mostly supports the view that subtle changes at the cellular and subcellular level, and in synaptic connectivity play major roles in the age-related decline in cognitive performance.

2.1.2. Age-related changes in zebrafish perceptual and cognitive performance

Zebrafish display a rich repertoire of behaviors, which depends on perceptual and cognitive processes [41]. As in other vertebrates, zebrafish have basic sensory systems and pathways for low-level sensory processing. For instance, the basic components and pathways of zebrafish visual system and visual processing hierarchy are similar to those commonly found in other species [42]. Of particular note, zebrafish can discriminate visual objects differing in color, shape and motion direction [43, 44]. There is a growing interest to assess motion perception acuity of zebrafish through optomotor responses or eye movements. Recent studies have shown that zebrafish perceive first- and second-order motions and also experience motion adaptations and illusions which are even thought to be seen only by humans [45, 46]. In addition, motion acuity and contrast sensitivity function have been found to be qualitatively similar to those of humans [47, 48]. These findings support the view that zebrafish visual system and perception (in particular, motion perception) rely on similar principles commonly found in humans. On the other hand, although there are studies comparing larval responses to visual stimulation (e.g., motion) with that of adult zebrafish [49], there is almost no systematic investigation on perceptual changes during aging. Future studies examining age-related changes in zebrafish perception and perceptual acuity will be informative in this respect and are currently being performed in our laboratory. Preliminary data suggest there are subtle age-related differences that are gender-dependent [50].

Behavioral studies also support the notion that zebrafish provide a promising model of cognitive functioning [51, 52]. It has been found that zebrafish have both simple (e.g., habituation, dishabituation and sensitization) and relatively complex forms of learning (e.g., associative and spatial learning). They also displayed good performance on tasks requiring either short-term (e.g., object recognition) or long-term (e.g., avoidance) memory. More importantly, the age-related cognitive declines have been shown by behavioral studies on learning and memory. For instance, zebrafish exhibit decreased performance with age on tasks relevant to associative learning and also show defects in spatial learning and avoidance with a distinct onset throughout aging [53, 54, 14]. In general, research on zebrafish will provide insight into potential neurobiological changes that would allow for application of interventions that would alter their course and possible translation to human populations.

2.2. Neurobiological changes

Behavioral and cognitive alterations that occur during normal aging are one of the most explored areas in the aging research since these are the manifestations of the aging itself, not to be confused with those related to pathologies. Their biological underpinnings, on the other hand, help us understand not only the cellular and synaptic mechanisms that play a role in aging, but also the very essence of the biology of behavior. To be able to understand the causes of cognitive decline that accompanies aging, for instance, we need to take all the biological components and their interaction with the environment into consideration. Previously, we described changes in cognitive processing that occur during aging and now we will focus on the neurobiological factors related to the hallmarks of aging [55] that likely underlie the changes in behavior and cognition (**Table 1**).

2.2.1. Epigenetic alterations and differential gene expression

Epigenetic mechanisms refer to structural or chemical modifications in RNA, DNA, and proteins without altering their primary sequence. These modifications play critical roles in major cellular processes such as regulation of gene expression, DNA replication, and cell cycle. Dynamic methylation/demethylation and acetylation/deacetylation events regulate structure of DNA and function of proteins, which consequently affect gene expression levels. DNA methylation at CpG dinucleotides, for example, is commonly associated with decreased DNA accessibility and turning genes off, although there are exceptions [56]. In contrast, histone acetylation generally results in an increase in gene expression [57]. The epigenetics of aging has been studied extensively, and researchers came up with a term, epigenetic drift, to define

Types of changes	Observations	Phenotypes	References
Epigenetic	Decrease in global DNA methylation	Disruption of gene expression and cellular differentiation	[59]
Gene expression	Decrease in expression of IGF signaling-related genes	Increase in lifespan	[16]
	Increase in *smurf2* expression	Replicative senescence	[16]
	Increase in *hsp1* expression	Impaired proteostasis	[11]
	Decrease in *tert* gene expression	Telomere shortening	[66]
Proteastasis	Decrease in Hsp70 levels	Impaired cellular stress response and proteostasis	[11]
	Increase in SOD2 activity		[64]
	Increased levels of lipofuscin		[65]
Genomic	DNA fragmentation	Senescence and/or cell loss	[60]
	Elevated apoptosis		[60]
Telomere attrition	Significant increase in the loss of telomere repeats	Telomere shortening	[66]
	Decrease in telomerase activity		[66]
Cellular and synaptic	Decrease in neurogenesis	Cognitive decline and altered behavior	[67, 68]
	Impaired oligodendrogenesis		[67]
	Altered balance in excitatory/inhibitory transmission		[75]

Table 1. Summary of age-related neurobiological changes.

age-related alterations in epigenetic patterns [58]. One well established epigenetic marker of aging is gradual decrease in global DNA methylation [59]. This decrease has been reported in humans, and other species including zebrafish. In zebrafish, while embryonic genome is highly methylated, gradual hypomethylation is observed throughout zebrafish lifespan [60]. These hypomethylation events are particularly observed at CpG islands, where several CpG dinucleotides cluster at regions typically involved with transcription regulation [60].

In addition to the previously mentioned global methylation events, differential gene expression, that could result in changes in cellular and synaptic functioning have been documented in the aging brain. In our study characterizing gene expression changes in the brains of young and old, male and female zebrafish, it was show that there are over 200 differentially expressed genes that are involved in cell differentiation, growth, neurogenesis, and brain and nervous system development [16]. For example, detailed analysis showed that expression of insulin-like growth factor (IGF) signaling-related genes including, igf1, igf2bp3, and igfbp2a, which are related to cell growth, significantly decreases in the brains of old zebrafish. In contrast, SMAD specific E3 ubiquitin protein ligase 2 (smurf2) expression, which is implicated in replicative senescence, is higher in old zebrafish compared to young [16]. Taken together, these data indicate potential differences in cellular and synaptic functioning.

2.2.2. Impaired proteostasis

Maintenance of protein homeostasis (a.k.a. proteostasis) is crucial for a cell's response to stress. Aging cells are exposed to increasing levels of stress, and there is even more need for protein quality control in these cells to keep the proteostatic balance for survival. When the networks that regulates protein synthesis, folding, and clearance malfunction due to age-related accumulation of cellular damage, proteostasis becomes impaired, making the proteome more vulnerable to cellular stress. In most organisms, gradual loss of proteostasis is associated with aging, yet there are some long-lived organisms that have more stable proteomes, suggesting an evidence for the importance of a stable proteome for longer lifespan [61].

Proteostasis networks involve chaperone proteins and two proteolytic systems: ubiquitin-proteasome and lysosome-autophagy systems. Chaperones guide newly synthesized proteins through processes that fold, transport and target for degradation [62]. The fate of unfolded proteins is collectively decided by all the proteolytic components mentioned before, but which proteolytic pathway to be followed depends mainly on chaperones. For example, heat shock protein 70 (Hsp70) family is comprised of several chaperones that are involved in stabilization of the correctly folded proteins and targeting proteins for degradation. In zebrafish, Hsp70 protein levels decrease with age, while hsp1 mRNA levels are increased in aged zebrafish, possibly to compensate for decreased ability of Hsp70 to function [11].

Activities of chaperone proteins are under the influence of age-related cellular changes such as dysregulation of cellular energetics. Reduced mitochondrial function, for example, impairs energy metabolism and limit the bioavailability of ATP. Aging cells respond to this reduction in ATP levels by switching to ATP-independent chaperones, as shown in aging human brain [63]. To overcome the effects of impaired energy metabolism and its harmful endproducts, cells respond in various ways. For example, increase in the major antioxidant enzyme in mitochondria, superoxide dismutase 2 (SOD2) activity has been reported in the aged zebrafish, although not in the brain tissue. SOD2 activity, on the other hand, increases between 3 and 12 months of age, indicating high metabolism, but decreases after 18 months, and even further decrease was observed in older zebrafish [64].

Dysregulation in lysosomal degradation may be a contributing factor to age-related cognitive dysfunction. For example, increased levels of lipofuscin, a non-degradable end product of lysosomal digestion, and oxidized proteins have been observed in lateral and medial pallial areas of the zebrafish brain at the age of 2 years as compared to 12 month-old animals [65]. Cognitive abilities related to memory are impaired in these fish starting from the age of 18 months, which is after the levels of these cellular byproducts started to increase. This enhanced oxidative stress, as observed by increases in lipofuscin and oxidized proteins, may be causing cognitive impairments in the aging zebrafish after reaching a certain threshold level of accumulation.

2.2.3. Genomic instability and telomere attrition

In the aging brain, as well as the whole body, there is an increase in genomic instability and telomere attrition. Mild DNA fragmentation, which is a biomarker of genomic instability, has been shown in young fish but the levels of fragmentation significantly increase after 12 months in various zebrafish tissues including brain. These changes in genomic instability in the aging zebrafish result in elevated apoptosis, which could reflect an increased need for removal of senescent or damaged cells [60]. The shortening of telomeres due to loss of repeats with each replication event, is another biomarker of aging. The mean telomere length of zebrafish decreases from the young adult stage to the older adult stage, with significant decreases in telomere length occurring after 18 months. This contrasts with the period before young adulthood, in which telomere length has been shown to increase. The decrease in telomere length is occurring after changes in the telomerase enzyme activity and expression, which starts to decrease in the eye and brain at the age of 6–12 months [66]. Research from young zebrafish with mutations in the gene encoding a telomere repeat binding factor 2 ($terfa^{hi3678/hi3678}$) support the role of proper telomere functioning as it relates to cellular homeostasis. These $terfa^{hi3678/hi3678}$ mutants die at early stages of development due to severe telomere shortening, which leads to premature retinal neurodegeneration, increased cellular senescence in the brain and spinal cord, which is shown by high SA-b-gal activity, and smaller eyes and head compared with wild type [18]. Heterozygotes are viable but have a shorter lifespan than their wild type counterparts [18]. Taken together changes in genomic instability and telomere length will contribute to cellular homeostasis.

2.2.4. Changes in cellular and synaptic capabilities

Both the mammalian and zebrafish brain have the capacity for cellular proliferation, although it is more limited in mammals. Changes in cellular proliferation may affect cognitive processing in the aged individual, and thus, zebrafish provide a good model for understanding mechanisms that regulate cell turnover in the brain. It has been observed by our research group and others that in both young and old adult zebrafish, neurogenesis is observed in the telencephalon, however, there is a significant decline in old adults [67, 68]. This decrease is attributed to the lengthening of cell cycle and decrease in the number of radial glia cells with self-renewal capability. In addition to this stem cell exhaustion, oligodendrogenesis has been shown to be impaired within the telencephalic parenchyma [67]. Therefore, similar to mammals, there is an age-related decrease in neural proliferation.

Loss of synaptic integrity may also play a major role in the age-related cognitive decline. Glutamate receptors, in particular, the N-methyl-D-aspartate (NMDA) receptor, have been implicated in age-related cognitive decline [69–71]. Decreases in the NMDA- and α-amino-3-hydroxy-5-methyl-4-isoxazolepropionic acid (AMPA)-type glutamate receptor levels have been shown to cause learning and memory impairments, whereas the increase leads to memory enhancement [72, 73]. While glutamate receptors remain a key molecular target that might be contributing to the age-related cognitive decline, it is important to understand changes in the excitatory/inhibitory balance and other neurotransmitter systems that might alter synaptic function with age. Compared to wild-types, mutants with impaired acetylcholinesterase function had better performance in spatial learning, entrainment and increased rate of learning [74, 14]. These findings suggest that cholinergic signaling may play a role in the age-related cognitive decline. Finally, we have begun to examine synaptic integrity in zebrafish brains. In a recent study, we examined changes in key synaptic proteins that reflect potential differences in both excitatory and inhibitory synapses across lifespan in young and old male and female zebrafish brains [75]. Our results show that the excitatory/inhibitory balance is altered differently in the brains of male and female zebrafish. These data indicate that there are age-related alterations in synaptic integrity that are gender-dependent and may contribute to cognitive decline.

3. Zebrafish models that delay or accelerate the aging process

Here, we will review several genetic and non-genetic interventions which are proposed to extend lifespan and healthspan, and discuss their use in zebrafish models in the context of aging.

3.1. Dietary restriction

Dietary restriction has been shown to have beneficial effects on both cognitive aging and the associated neurobiological changes in the aging brain. This has been shown in humans [76] and animal models [77–80]. Our research group proposed utilizing a dietary manipulation such as caloric restriction (CR) to alter the aging process in zebrafish. To date, only a few studies apart from our own have utilized a true CR in fish [68]. The previously published dietary restriction studies that have been performed in zebrafish did not reduce daily caloric intake but rather the fish were not given any food for an extended period, which would be considered as a starvation study [81, 82] and none had been done in aged animals or with regards to gender. Thus, designing the appropriate dietary intervention needed to be based on a wealth of literature that has accumulated from studies utilizing different animal models.

Initially, our research group established a protocol designed as a daily reduction in caloric intake in zebrafish. For this task, cohorts of both young and old fish raised in the zebrafish facility were moved to round glass aquaria (**Figure 1a** and **b**). Fish were fed individually in 600 mL beakers during the weekdays (**Figure 1c** and **Table 2**). On the weekend, fish were fed similar amounts of food in the housing aquaria. It should be noted that the animals were not housed continually in the beakers since zebrafish are highly social [83, 84], and continuous social isolation would increase their stress levels [85]. Since the effects of CR are thought to be modulated through the target of rapamycin (TOR) pathway [86], we aimed to test whether we could mimic the effects of CR with rapamycin treatment, a TOR inhibitor. Rapamycin is a macrocyclic compound produced by bacterium *Streptomyces hygroscopicus* and approved for patient use by the Food and Drug Administration (FDA, USA) [87]. The rapamycin group was treated daily with 100 nM rapamycin dissolved in DMSO. The fish in all treatment groups

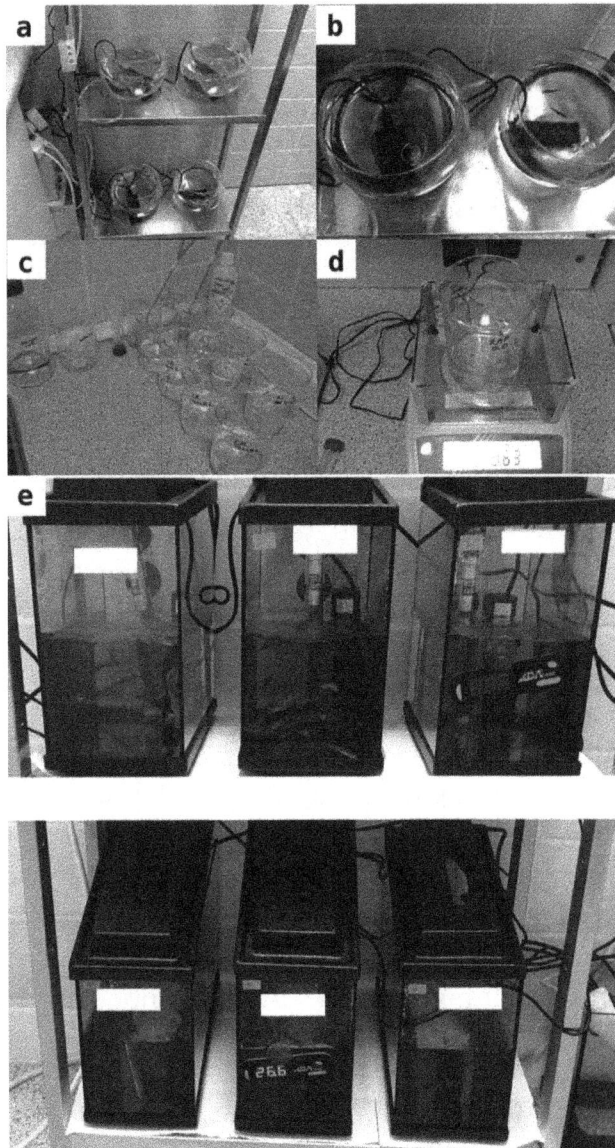

Figure 1. Aquaria set-up for CR and rapamycin treatment (a–d), and IF and rapamycin treatment (e) experiments.

	Monday	Tuesday	Wednesday	Thursday	Friday	Saturday	Sunday
Ad libitum (control for CR and rapamycin-treated groups)	20 mg*	20 mg*	20 mg*	20 mg*	20 mg*	20 mg*	20 mg*
Rapamycin treatment	20 mg*	20 mg*	20 mg*	20 mg*	20 mg*	20 mg*	20 mg*
Caloric restriction	1 mg*	1 mg*	1 mg*	1 mg*	1 mg*	1 mg*	1 mg*
Ad libitum (control for IF and rapamycin-treated groups)	90 mg food** and artemia†	90 mg food**	90 mg food** and artemia†	90 mg food**	90 mg food** and artemia†	90 mg food**	90 mg food**
Rapamycin treatment	90 mg food** and artemia†	90 mg food**	90 mg food** and artemia†	90 mg food**	90 mg food** and artemia†	90 mg food**	90 mg food**
Intermittent feeding	45 mg food**		45 mg food** and artemia†		45 mg food**		45 mg food**
Overfed	180 mg food** and artemia**	180 mg food** and artemia**	180 mg food** and artemia**	180 mg food** and artemia**	180 mg food** and artemia**	180 mg food** and artemia**	180 mg food** and artemia**

*Food per fish per day.

**Twice a day.

†Once a day.

Table 2. Different feeding paradigms applied by our research group to study the effects of dietary restriction (i.e. CR and IF) or overfeeding compared to drug treatment (i.e. rapamycin-treated) on zebrafish aging.

were weighed individually in beakers throughout the experiment (**Figure 1d**). The results demonstrated small losses in body weight in all groups (**Figure 2a**). While these data indicated that CR caused a significant weight loss, a better and more efficient CR protocol was needed since all the fish lost weight, and this likely indicates that all animals were under some stress.

CR can be performed as a daily reduction in caloric intake or as an every-other-day feeding regimen, also known as an intermittent fasting (IF) paradigm. Research has shown that the effects on body weight and markers related to cellular and synaptic plasticity in the brain are not different for these two paradigms [88]. In our recent study [68], we utilized an IF regimen that included dry flakes and artemia (**Table 2**). An IF paradigm would not disrupt any social hierarchy of the fish or cause any unnecessary social isolation or netting stress in the fish since the animals would be kept in their home tanks. Artemia is not only an extra source of protein but also provides environmental enrichment to the fish [89]. The diet continued for 10 weeks and the results indicated that while there was no prevention of an age-related decline in newly born neurons, IF treatment stabilized an age-related decline in telomere shortening [68]. Thus, some of the beneficial effects of dietary restriction maybe done through subtly altering the cell cycle dynamics.

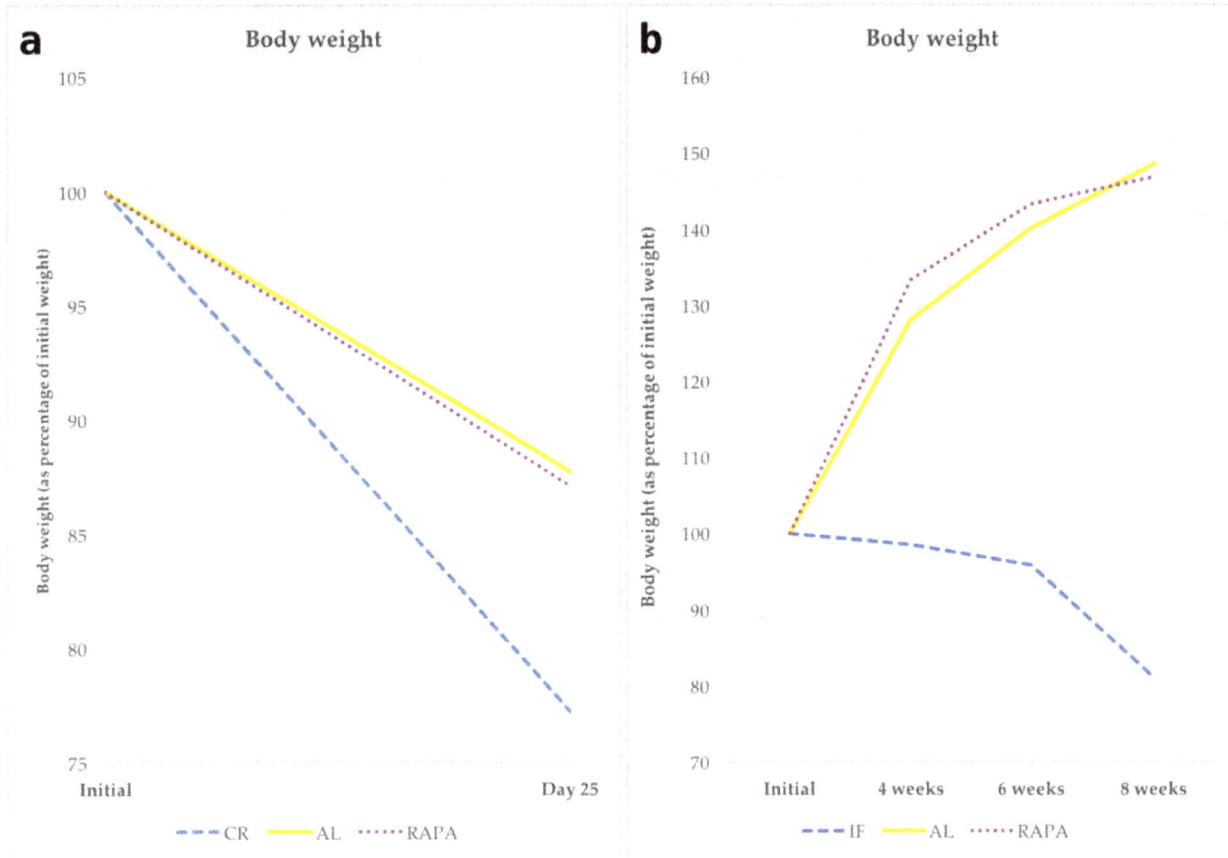

Figure 2. (a) Final body weight on day 25 as percentage of the initial body weight. AL and RAPA animals lost 12 and 13% of their initial weight, respectively, whereas CR animals lost 23%. (b) The effect of IF and rapamycin treatment on body weight of animals. Body weights at 4, 6 and 8 weeks were compared to the average weight at the beginning of the experiment. While AL and RAPA animals showed a very similar pattern of weight gain throughout 8-week experiment, average weight of IF animals was first stabilized, then after 4 weeks they continuously lost weight, with the largest decline occurring between 6 and 8 weeks.

Our group has extended this paradigm to include a rapamycin-treated group to test whether rapamycin could mimic CR's effects [90], and careful design was necessary since the fish remained continuously in the tanks (**Figure 1e**) while receiving the drug treatment and the half-life of the drug needed to be considered. According to the literature, half-life of rapamycin is 3 days [87] so at least half and at most three-quarters of the tank water was replaced with fresh water supplemented with 100 nM rapamycin every 3 days to keep the active drug levels consistent (**Table 2**). We applied the same protocol for water replacement for *ad libitum* (AL) and IF groups without adding rapamycin, since the change of aquaria water also permitted us to keep nitrate levels low and pH levels stable. Our preliminary data demonstrated that body weight decreased significantly after 6–8 weeks of IF treatment but was not different in the AL or rapamycin-treated animals [91] (**Figure 2b**). These data suggest that an IF paradigm in zebrafish can significantly reduce body weight in young and old animals, and the effects of IF can be studied in conjunction with a drug-treated group. Current analysis is being performed as to whether these treatments alter the course of some of the neurobiological alterations that were referred to in Section 2.2. For example, CR may regulate DNA methylation at individual loci by increasing the activities and/or expression levels of certain DNA methyltransferase (DNMT) enzymes in different animal models [92, 93]. In addition to the regulation of DNMTs, histone acetylation and deacetylation are affected by CR [94]. Induction of neuroprotection-related gene expression patterns by CR, such as upregulation of *mir-98-3p* in the cerebral cortex of rats have also been shown [95]. Also, CR regulates synaptic proteins in the aging brain and delays cognitive declines in memory [96]. Therefore, the durations of CR or IF alongside potential mimetics of dietary restriction are important to examine in zebrafish for potential translation to human studies.

3.2. Overfeeding

Contrary to CR and IF paradigms that decelerate the aging process, overfeeding has been proposed to accelerate aging. For example, the effects of obesity on behavioral and neuro-biological changes have been investigated in humans [76] and animal models [97]. Zebrafish, like humans, will overeat if exposed to large amounts of food [98]. Although studies have examined the effects of overfeeding in zebrafish [99–101], few studies have utilized these animals as a model to study obesity, and to a lesser extent to study the effects of brain aging. A recent study demonstrated that a life-long high caloric diet in adult zebrafish caused significant increases in body weight, and eventually obesity, along with high cortisol levels and decreased rate of neurogenesis which were similar to the observations on older fish [101].

In order to compare our dietary restriction and drug manipulations with overfeeding, our research group decided to test whether short-term overfeeding in fish will lead to premature neurobiological changes in the aging brain in young adults and even further accelerated brain aging in older fish. For this paradigm, we increased the amount of dry food per feeding to two times the normal quantity and supplied artemia every day instead of three times a week. These overfed fish will be compared back to our regular AL- and IF-treated groups (**Table 2**). Our initial observations indicate that the animals exposed to overfeeding have an increase in body weight in comparison to AL-fed and IF animals. This type of feeding paradigm, along with examining the effects of short-term dietary restriction, has important implications for possible interventions and translational studies for humans.

3.3. Mutant models

Accelerated or decelerated brain aging may be affected by environmental manipulations such as overfeeding or dietary restriction. However, it is difficult to establish the link between direct cellular and molecular mechanisms affecting the behavioral or neurobiological changes upon such dietary manipulations. As was mentioned earlier, one of the advantages of the zebrafish animal model is the ease of manipulating genes. This can be done with forward genetics screening where one can identify mutated genes responsible for a particular disease phenotype, or it is possible to introduce a mutation in a specific gene of interest by using reverse genetic screens such as the transposon-mediated insertional mutagenesis (Tol2) or CRISPR-Cas9 system. Using inducible tissue- or cell-specific promotors a gene could be expressed at anatomically distinct areas in the brain, and thus a more refined understanding of the gene function. Taken together, all of these available genetic tools and approaches allow zebrafish researchers to perform large-scale genetic screens in order to understand the function of corresponding genes during development and adulthood.

Identifying genes which play role in decelerated and accelerated aging models are critically important. One such example, which was obtained from a chemical screen, is a fish line with a specific mutation in the acetylcholinesterase gene ($ache^{sb55}$) that results in a loss-of-function of this enzyme and an increase in the levels of acetylcholine [102]. In the normal aging brain, there is a loss of acetylcholine that is thought to contribute to age-related cognitive decline [103], so the fish with increased acetylcholine levels should have delayed brain aging and cognitive deficits. While homozygous mutants die in very early developmental stages, interestingly, mutants with a heterozygous mutation will develop and live until late adulthood. Furthermore, Yu et al. demonstrated that age-related spatial learning deficits are delayed in these animals, which is consistent with the hypothesis that acetylcholine levels affect age-related cognitive function [14]. Thus, these data suggest that preventing age-related declines in acetylcholine can alter the pattern of age-related phenotypes.

In order to understand the direct cellular mechanism of dietary restriction, we are currently examining the effects of altering the TOR signaling pathway in zebrafish. We have two models with which we are currently working. The first is a zebrafish line with a knockout of the *tor* gene using the Tol2 system that resulted in a mutated TOR [104]. In the second we are creating transgenic models of zebrafish using the Tol2 system that will express one of the following types of TOR, (1) an overactivated mouse TOR complex, (2) an overactivated mouse TOR complex that is rapamycin resistant, (3) a reduced or an inactive mouse TOR complex, or (4) a reduced or an inactive mouse TOR complex that that is rapamycin resistant. This will allow us to examine the effects of dietary restriction and rapamycin separately. While these data regarding $ache^{sb55}$ mutants and the data from potential TOR transgenics are interesting, a confound still exists that needs to be addressed, which is that in both cases the mutations are active throughout the entire life of the animal, including early developmental time periods. Thus, future studies need to be directed at creating tissue- or cell-specific lines that are inducible, which can be easily be done in zebrafish.

3.4. Transient gene knockdowns and overexpression in adults and embryos

One of the useful genetic tools available for use in zebrafish is injections of both morpholino antisense oligonucleotides and *in vitro* transcribed capped mRNA. The morpholinos create a transient knockdown of a specific gene of interest and are an essential tool for understanding the specific role of that gene. Microinjecting morpholino antisense oligonucleotides targets and blocks access of the cellular components to specific RNAs and by doing so, can prevent translation, splicing, or ribozyme activity depending on their design, and results in a loss of a functional protein [105, 106]. Microinjection of *in vitro* transcribed capped mRNA, on the other hand, serves as a powerful tool for overexpressing a gene of interest to understand its function *in vivo*, which is in contrast to the knockdown approach with morpholinos [107–109]. From a mechanistic point of view, morpholino antisense oligonucleotides can be designed in several ways, either translation-blocking or splice-blocking [105, 106]. As its name implies, a translation-blocking morpholino oligonucleotide binds to its target RNA and blocks the sites where the translation initiation machinery accesses the target RNA molecules. The splice-blocking morpholino oligonucleotide inhibits the splicing of pre-mRNA. Besides these traditional types of morpholino, there is also the vivo-morpholino, which is a promising tool for transient gene silencing in adult animals and cell culture [110–115]. These vivo-morpholino experiments can also be compared to injections of *in vitro* transcribed capped mRNA into the adult brain to overexpress a gene of interest. Taken together these techniques allow for silencing or overexpressing a gene of interest in both embryos and adults.

Injections of both morpholino antisense oligonucleotides and *in vitro* transcribed capped mRNA can be easily done in embryos and adults. The injections into embryos are done in the 1–4 cell stage. A method utilized for both vivo-morpholino and *in vitro* capped mRNA injections in the adult brain is performed by cerebroventricular microinjection (CVMI). The CVMI technique was used by Kizil et al. [116, 117] to investigate the effects of overproduction of amyloid protein on cellular, molecular and functional aspects of the adult zebrafish brain, and has been shown to be a useful method for inactivation studies in zebrafish and other vertebrates [118–123]. These techniques in the adults hold promise to provide an understanding of the role of a gene of interest without the possible confounding effects of early developmental changes on the future adult animal. Moreover, the embryo injections not only give insight into the developmental role of a gene of interest but might also provide a platform for an *in vivo* cell culture model for examining the cellular mechanisms of aging and aging-related interventions.

4. Conclusions

Based on the similarities to the aging process in humans, the zebrafish is emerging as a promising model to understand the mechanisms that lead to age-related behavioral and neurobiological alterations. Additionally, there are many genetic tools such as mutagenesis protocols and transient knockdown/overexpression approaches that can easily be applied to the zebrafish to intervene in this aging process, in addition to the many potential environmental manipulations. Moreover, it is possible to study different behavioral paradigms such as visual motion processing and spatial learning and memory abilities that change with increasing age. In these behaviorally-characterized animals it will be feasible to measure the expression levels of neurobiological markers to understand their functional role in cognition, and determine the success of potential interventions. The zebrafish is also a powerful model for the use of drug screening and provides information about promising therapeutics that can be eventually translated to human populations. All in all, this gerontological model will be very useful for unlocking the secrets of the aging brain and potentially helping to uncover anti-aging therapies in order to restore brains back to their youthful capacity.

Acknowledgements

The current studies by our group mentioned in this chapter are being supported by grants (214S236 and 215S701) under the 1001 project scheme from the Scientific and Technological Research Council of Turkey (TÜBİTAK). The initial caloric restriction/rapamycin study was supported by an Installation Grant from the European Molecular Biology Organization (EMBO).

Conflict of interest

The authors declare no conflict of interest.

Notes/thanks/other declarations

The authors would like to thank Tülay Arayıcı for providing excellent technical assistance for all the experiments mentioned in this chapter that were conducted in the Department of Molecular Biology and Genetics Zebrafish Facility in Bilkent University. Additionally, the authors thank Dr. Ayça Arslan-Ergül, Göksemin Fatma Şengül, and Narin Ilgım Ardıç for help with feeding and tissue harvesting in the experiments. Finally, the authors thank Elif Karoğlu for helpful discussion with regards to the information for the design of the overfeeding experiments.

Author details

Dilan Celebi-Birand[1,2,3], Begun Erbaba[1,2,3], Ahmet Tugrul Ozdemir[3,4], Hulusi Kafaligonul[1,3,5] and Michelle Adams[1,2,3,6]*

*Address all correspondence to: michelle@bilkent.edu.tr

1 Interdisciplinary Graduate Program in Neuroscience, Aysel Sabuncu Brain Research Center, Bilkent University, Turkey

2 UNAM-National Nanotechnology Research Center and Institute of Materials Science and Nanotechnology, Bilkent University, Turkey

3 Department of Molecular Biology and Genetics Zebrafish Facility, Bilkent University, Turkey

4 Division of Cognitive Neurobiology, Center for Brain Research, Medical University of Vienna, Vienna, Austria

5 UMRAM-National Magnetic Resonance Research Center, Bilkent University, Turkey

6 Department of Psychology, Bilkent University, Turkey

References

[1] Martin GM. Modalities of gene action predicted by the classical evolutionary biological theory of aging. Annals of the New York Academy of Sciences. 2007;**1100**:14-20

[2] Martin GM, Bergman A, Barzilai N. Genetic determinants of human health span and life span: Progress and new opportunities. PLoS Genetics. 2007;**3**:e125

[3] Tissenbaum HA. Using *C. elegans* for aging research. Invertebrate Reproduction & Development. 2014;**2014**:59-63

[4] He Y, Jasper H. Studying aging in Drosophila. Methods. 2014;**68**:129-133

[5] Ackert-Bicknell CL, Anderson LC, Sheehan S, Hill WG, Chang B, Churchill GA, et al. Aging research using mouse models. Current Protocols in Mouse Biology. 2015;**5**:95-133

[6] Herbig U, Ferreira M, Condel L, Carey D, Sedivy JM. Cellular senescence in aging primates. Science. 2006;**311**:1257

[7] Hoffman KL, McNaughton BL. Coordinated reactivation of distributed memory traces in primate neocortex. Science. 2002;**297**:2070-2073

[8] Francis PJ, Appukuttan B, Simmons E, Landauer N, Stoddard J, Hamon S, Ott J, Ferguson B, Klein M, Stout JT, Neuringer M. Rhesus monkeys and humans share common susceptibility genes for age-related macular disease. Human Molecular Genetics. 2008;**17**:2673-2680

[9] Peters A. Structural changes in the normally aging cerebral cortex of primates. Progress in Brain Research. 2002;**136**:455-465

[10] Kishi S, Uchiyama J, Baughman AM, Goto T, Lin MC, Tsai SB. The zebrafish as a vertebrate model of functional aging and very gradual senescence. Experimental Gerontology. 2003;**38**:777-786

[11] Keller ET, Murtha JM. The use of mature zebrafish (*Danio rerio*) as a model for human aging and disease. Comparative Biochemistry and Physiology Part C: Toxicology & Pharmacology. 2004;**138**:335-341

[12] Gerhard GS, Kauffman EJ, Wang X, Stewart R, Moore JL, Kasales CJ, Demidenko E, Cheng KC. Life spans and senescent phenotypes in two strains of zebrafish (*Danio rerio*). Experimental Gerontology. 2002;**37**:1055-1068

[13] Zhdanova IV, Yu L, Lopez-Patino M, Shang E, Kishi S, Guelin E. Aging of the circadian system in zebrafish and the effects of melatonin on sleep and cognitive performance. Brain Research Bulletin. 2008;**75**:433-441

[14] Yu L, Tucci V, Kishi S, Zhdanova IV. Cognitive aging in zebrafish. PLoS One. 2006;**1**:e14

[15] Tsai SB, Tucci V, Uchiyama J, Fabian NJ, Lin MC, Bayliss PE, Neuberg DS, Zhdanova IV, Kishi S. Differential effects of genotoxic stress on both concurrent body growth and gradual senescence in the adult zebrafish. Aging Cell. 2007;**6**:209-224

[16] Arslan-Ergul A, Adams MM. Gene expression changes in aging zebrafish (*Danio rerio*) brains are sexually dimorphic. BMC Neuroscience. 2014;**15**:29

[17] Munzel EJ, Becker CG, Becker T, Williams A. Zebrafish regenerate full thickness optic nerve myelin after demyelination, but this fails with increasing age. Acta Neuropathologica Communications. 2014;**2**:77

[18] Kishi S, Bayliss PE, Uchiyama J, Koshimizu E, Qi J, Nanjappa P, Imamura S, Islam A, Neuberg D, Amsterdam A, Roberts TM. The identification of zebrafish mutants showing alterations in senescence-associated biomarkers. PLoS Genetics. 2008;**4**:e1000152

[19] Zhdanova IV. Melatonin as a hypnotic: Pro. Sleep Medicine Reviews. 2005;**9**:51-65

[20] Zhdanova IV, Wang SY, Leclair OU, Danilova NP. Melatonin promotes sleep-like state in zebrafish. Brain Research. 2001;**903**:263-268

[21] Williams FE, White D, Messer WS. A simple spatial alternation task for assessing memory function in zebrafish. Behavioural Processes. 2002;**58**:125-132

[22] Levin ED, Chen E. Nicotinic involvement in memory function in zebrafish. Neurotoxicology and Teratology. 2004;**26**:731-735

[23] Levin ED, Limpuangthip J, Rachakonda T, Peterson M. Timing of nicotine effects on learning in zebrafish. Psychopharmacology. 2005;**184**:547-552

[24] Levin ED, Cerutti DT. Behavioral neuroscience of zebrafish. In: Buccafusco JJ, editor. Methods of Behavior Analysis in Neuroscience. New York: CRC Press; 2008. pp. 293-310

[25] Wullimann MF, Rupp B, Reichert H. Neuroanatomy of the Zebrafish Brain: A Topological Atlas. Basel: Birkhäuser Verlag; 1996

[26] Vargas R, Jóhannesdóttir IÞ, Sigurgeirsson B, Þorsteinsson H, Karlsson KÆ. The zebrafish brain in research and teaching: A simple in vivo and in vitro model for the study of spontaneous neural activity. Advances in Physiology Education. 2011;**35**:188-196

[27] Nam RH, Kim W, Lee CJ. NMDA receptor-dependent long-term potentiation in the telencephalon of the zebrafish. Neuroscience Letters. 2004;**370**:248-251

[28] Ganz J, Kroehne V, Freudenreich D, Machate A, Geffarth M, Braasch I, et al. Subdivisions of the adult zebrafish pallium based on molecular marker analysis. F1000Research. 2014;**3**:308

[29] Wullimann MF, Mueller T. Teleostean and mammalian forebrains contrasted: Evidence from genes to behavior. The Journal of Comparative Neurology. 2004;**475**:143-162

[30] Owsley C. Aging and vision. Vision Research. 2011;**51**:1610-1622

[31] Wang Z, Yao Z, Yuan N, Liang Z, Li G, Zhou Y. Declined contrast sensitivity of neurons along the visual pathway in aging cats. Frontiers in Aging Neuroscience. 2014;**6**:163

[32] Roudaia E, Bennett PJ, Sekuler AB, Pilz KS. Spatiotemporal properties of apparent motion perception and aging. Journal of Vision. 2010;**10**(5):1-15

[33] Snowden RJ, Kavanagh E. Motion perception in the ageing visual system: Minimum motion, motion coherence, and speed discrimination thresholds. Perception. 2006;**35**:9-24

[34] Wang H, Xie X, Li X, Chen B, Zhou Y. Functional degradation of visual cortical cells in aged rats. Brain Research. 2006;**1122**:93-98

[35] Yang Y, Liang Z, Li G, Wang Y, Zhou Y, Leventhal AG. Aging affects contrast response functions and adaptation of middle temporal visual area neurons in rhesus monkeys. Neuroscience. 2008;**156**:748-757

[36] Grady CL. Cognitive neuroscience of aging. Annals of the New York Academy of Sciences. 2008;**1124**:127-144

[37] Salthouse TA. The role of processing resources in cognitive aging. In: Howe ML, Brainerd CJ, editors. Cognitive Development in Adulthood: Progress in Cognitive Development Research. New York: Springer; 1988. pp. 185-239

[38] Salthouse TA. Aging and measures of processing speed. Biological Psychology. 2000;**54**:35-54

[39] Gold P, McGauch JL. Changes in learning and memory during aging. In: Ordy J, editor. Advances in Behavioral Biology. New York: Plenum; 1975. p. 145-158

[40] Grady CL, Craik FIM. Changes in memory processing with age. Current Opinion in Neurobiology. 2000;**10**:224-231

[41] Meshalkina DA, Kizlyk MN, Kysil EV, Collier AD, Echevarria DJ, Abreu MS, et al. Understanding zebrafish cognition. Behavioural Processes. 2017;**141**:229-241

[42] Bilotta J, Saszik S. The zebrafish as a model visual system. International Journal of Developmental Neuroscience. 2001;**19**:621-629

[43] Oliveira J, Silveira M, Chacon D, Luchiari A. The zebrafish world of colors and shapes: Preference and discrimination. Zebrafish. 2015;**12**:166-173

[44] Orger MB, Smear MC, Anstis SM, Baier H. Perception of Fourier and non-Fourier motion by larval zebrafish. Nature Neuroscience. 2000;**3**:1128-1133

[45] Gori S, Agrillo C, Dadda M, Bisazza A. Do fish perceive illusory motion? Scientific Reports. 2014;**4**:6443

[46] Najafian M, Alerasool N, Moshtaghian J. The effect of motion aftereffect on optomotor response in larva and adult zebrafish. Neuroscience Letters. 2014;**559**:179-183

[47] Haug MF, Biehlmaier O, Mueller KP, Neuhauss SC. Visual acuity in larval zebrafish: Behavior and histology. Frontiers in Zoology. 2010;**7**:8

[48] Tappeiner C, Gerber S, Enzmann V, Balmer J, Jazwinska A, Tschopp M. Visual acuity and contrast sensitivity of adult zebrafish. Frontiers in Zoology. 2012;**9**:10

[49] Bak-Coleman J, Smith D, Coombs S. Going with, then against the flow: Evidence against the optomotor hypothesis of fish rheotaxis. Animal Behaviour. 2015;**107**:7-17

[50] Karaduman A, Kaya U, Karoglu ET, Ergul-Arslan A, Adams MM, Kafaligonul H. Motion direction discrimination during neural aging. Program No. 685.01/GG9 Neuroscience Meeting Planner. Washington, DC: Society for Neuroscience, Online; 2017

[51] Blaser R, Vira D. Experiments on learning in zebrafish (*Danio rerio*): A promising model of neurocognitive function. Neuroscience & Biobehavioral Reviews. 2014;**42**:224-231

[52] Gerlai R. Learning and memory in zebrafish (*Danio rerio*). Methods in Cell Biology The Zebrafish—Cellular and Developmental Biology, Part B Developmental Biology. 2016;**134**:551-586

[53] Brock AJ, Sudwarts A, Parker MO, Brennan CH. Zebrafish Behavioral models of ageing. In: Kalueff AV, editor. The Rights and Wrongs of Zebrafish: Behavioral Phenotyping of Zebrafish. Cham: Springer International Publishing; 2017. pp. 241-258

[54] Arey RN, Murphy CT. Conserved regulators of cognitive aging: From worms to humans. Behavioural Brain Research. 2017;**322**:299-310

[55] López-Otín C, Blasco MA, Partridge L, Serrano M, Kroemer G. The hallmarks of aging. Cell. 2013;**153**:1194-1217

[56] Sierra MI, Fernandez AF, Fraga MF. Epigenetics of aging. Current Genomics. 2015;**16**:435-440

[57] Kouzarides T. Chromatin modifications and their function. Cell. 2007;**128**:693-705

[58] Fraga MF, Agrelo R, Esteller M. Cross-talk between aging and cancer: The epigenetic language. Annals of the New York Academy of Sciences. 2007;**1100**:60-74

[59] Wilson VL, Smith RA, Ma S, Cutler RG. Genomic 5-methyldeoxycytidine decreases with age. The Journal of Biological Chemistry. 1987;**262**:9948-9951

[60] Shimoda N, Hirose K, Kaneto R, Izawa T, Yokoi H, Hashimoto N. No evidence for AID/MBD4-coupled DNA demethylation in zebrafish embryos. PLoS One. 2014;**9**:e114816

[61] Treaster SB, Ridgway ID, Richardson CA, Gaspar MB, Chaudhuri AR, Austad SN. Superior proteome stability in the longest lived animal. Age. 2013;**36**:9597

[62] Feldman DE, Frydman J. Protein folding in vivo: The importance of molecular chaperones. Current Opinion in Structural Biology. 2000;**10**:26-33

[63] Brehme M, Voisine C, Rolland T, Wachi S, Soper J, Zhu Y, et al. A chaperome subnetwork safeguards proteostasis in aging and neurodegenerative disease. Cell Reports. 2014;**9**:1135-1150

[64] Almaida-Pagán PF, Lucas-Sánchez A, Tocher DR. Changes in mitochondrial membrane composition and oxidative status during rapid growth, maturation and aging in zebrafish, *Danio rerio*. Biochimica et Biophysica Acta (BBA)—Molecular and Cell Biology of Lipids. 2014;**1841**:1003-1011

[65] Ruhl T, Jonas A, Seidel NI, Prinz N, Albayram O, Bilkei-Gorzo A, von der Emde G. Oxidation and cognitive impairment in the aging zebrafish. Gerontology. 2015;**62**:47-57

[66] Anchelin M, Murcia L, Alcaraz-Pérez F, García-Navarro EM, Cayuela ML. Behaviour of telomere and telomerase during aging and regeneration in zebrafish. PLoS One. 2011;**6**:e16955

[67] Edelmann K, Glashauser L, Sprungala S, Hesl B, Fritschle M, Ninkovic J, Godinho L, Chapouton P. Increased radial glia quiescence, decreased reactivation upon injury and unaltered neuroblast behaviour underlie decreased neurogenesis in the aging zebrafish telencephalon. Journal of Comparative Neurology. 2013;**521**:3099-3115

[68] Arslan-Ergul A, Erbaba B, Karoglu ET, Halim DO, Adams MM. Short-term dietary restriction in old zebrafish changes cell senescence mechanisms. Neuroscience. 2016;**334**:64-75

[69] Barnes CA. Normal aging: Regionally specific changes in hippocampal synaptic transmission. Trends in Neuroscience. 1994;**17**:13-18

[70] Morrison JH, Gazzaley AH. Age-related alterations of the N-methyl-D-aspartate receptor in the dentate gyrus. Molecular Psychiatry. 1996;**1**:356-358

[71] Morrison JH, Hof PR. Life and death of neurons in the aging brain. Science. 1997;**278**:412-419

[72] Tang YP, Shimizu E, Dube GR, Rampon C, Kerchner GA, Zhuo M. Genetic enhancement of learning and memory in mice. Nature. 1999;**401**:63-69

[73] Tsien JZ, Huerta PT, Tonegawa S. The essential role of hippocampal CA1 NMDA receptor-dependent synaptic plasticity in spatial memory. Cell. 1996;**87**:1327-1338

[74] Parker MO. Developmental role of acetylcholinesterase in impulse control in zebrafish. Frontiers in Behavioral Neuroscience. 2015;**9**:271

[75] Karoglu ET, Halim DO, Erkaya B, Altaytas F, Arslan-Ergul A, Konu O, Adams MM. Aging alters the molecular dynamics of synapses in a sexually dimorphic pattern in zebrafish (*Danio rerio*). Neurobiology of Aging. 2017;**54**:10-21

[76] Stillman CM, Weinstein AM, Marsland AL, Gianaros PJ, Erickson KI. Body–brain connections: The effects of obesity and behavioral interventions on neurocognitive aging. Frontiers in Aging Neuroscience. 2017;**9**:115

[77] Adams MM, Shi L, Linville MC, Forbes ME, Long AB, Bennett C, Newton IG, Carter CS, Sonntag WE, Riddle DR. Caloric restriction and age affect synaptic protein levels in hip-pocampal CA3 and spatial learning ability. Experimental Neurology. 2008;**211**:141-149

[78] Ingram DK, Weindruch R, Spangler EL, Freeman JR, Walford RL. Dietary restriction benefits learning and motor performance of aged mice. The Journals of Gerontology. 1987;**42**:78-81

[79] Markowska AL, Savonenko A. Retardation of cognitive aging by life-long diet restriction: Implications for genetic variance. Neurobiology of Aging. 2002;**23**:75-78

[80] Stewart J, Mitchell J, Kalant N. The effects of life-long food restriction on spatial memory in young and aged Fischer 344 rats measured in the eight-arm radial and the Morris water mazes. Neurobiology of Aging. 1989;**10**:669-675

[81] Novak CM, Jiang X, Wang C, Teske JA, Kotz CM, Levine JA. Caloric restriction and physical activity in zebrafish (*Danio rerio*). Neuroscience Letters. 2005;**383**:99-104

[82] Craig PM, Moon TW. Fasted zebrafish mimic genetic and physiological responses in mammals: A model for obesity and diabetes? Zebrafish. 2011;**8**:109-117

[83] Kishi S, Slack BE, Uchiyama J, Zhdanova IV. Zebrafish as a genetic model in biological and behavioral gerontology: Where development meets aging in vertebrates. Gerontology. 2009;**55**:430-441

[84] Lieschke GJ, Currie PD. Animal models of human disease: Zebrafish swim into view. Nature Reviews Genetics. 2007;**8**:353-367

[85] Pavlidis M, Digka N, Theodoridi A, Campo A, Barsakis K, Skouradakis G, Samaras A, Tsalafouta A. Husbandry of zebrafish, *Danio rerio*, and the cortisol stress response. Zebrafish. 2013;**10**:524-531

[86] Dogan S, Johannsen AC, Grande JP, Cleary MP. Effects of intermittent and chronic calorie restriction on mammalian target of rapamycin (mTOR) and IGF-I signaling pathways in mammary fat pad tissues and mammary tumors. Nutrition and Cancer. 2011;**63**:389-401

[87] FDA. DOSAGE AND ADMINISTRATION [Internet]. Available from: https://www.fda. gov/ohrms/dockets/ac/02/briefing/3832b1_03_FDA-RapamuneLabel.htm [Accessed: 2018-01-25]

[88] Murphy T, Dias GP, Thuret S. Effects of diet on brain plasticity in animal and human studies: Mind the gap. Neural Plasticity. 2014:1-32

[89] Varga Z. Aquaculture, husbandry, and shipping at the Zebrafish International Resource Center. Methods in Cell Biology The Zebrafish—Genetics, Genomics, and Transcriptomics. 2016;**135**:509-534

[90] Halloran J, Hussong SA, Burbank R, Podlutskaya N, Fischer KE, Sloane LB. Chronic inhibition of mammalian target of rapamycin by rapamycin modulates cognitive and non-cognitive components of behavior throughout lifespan in mice. Neuroscience. 2012;**223**:102-113

[91] Celebi-Birand ED, Sengul GF, Ardic NI, Kafaligonul H, Adams MM. Effects of Short-Term Caloric Restriction and Rapamycin Treatment on Cellular and Synaptic Components in Young and Old Zebrafish (*Danio rerio*). Program No. 663.15/K6 Neuroscience Meeting Planner. Washington, DC: Society for Neuroscience, Online; 2017

[92] Muñoz-Najar U, Sedivy JM. Epigenetic control of aging. Antioxidants & Redox Signaling. 2011;**14**:241-259

[93] Li Y, Daniel M, Tollefsbol TO. Epigenetic regulation of caloric restriction in aging. BMC Medicine. 2011;9:98

[94] Li Y, Liu L, Tollefsbol T. Glucose restriction can extend normal cell lifespan and impair precancerous cell growth through epigenetic control of *hTERT* and *p16* expression. FASEB Journal. 2010;24:1442-1453

[95] Wood SH, Dam SV, Craig T, Tacutu R, O'Toole A, Merry BJ, Magalhães JP. Transcriptome analysis in calorie-restricted rats implicates epigenetic and post-translational mechanisms in neuroprotection and aging. Genome Biology. 2015;16:285

[96] Shi L, Adams MM, Linville MC, Newton IG, Forbes ME, Long AB, Riddle DR, Brunso-Bechtold JK. Caloric restriction eliminates the aging-related decline in NMDA and AMPA receptor subunits in the rat hippocampus and induces homeostasis. Experimental Neurology. 2007;206:70-79

[97] Uranga RM, Bruce-Keller AJ, Morrison CD, Fernandez-Kim SO, Ebenezer PJ, Zhang L, et al. Intersection between metabolic dysfunction, high fat diet consumption, and brain aging. Journal of Neurochemistry. 2010;114:344-361

[98] Westerfield M. The Zebrafish Book. A Guide for the Laboratory Use of Zebrafish (*Danio rerio*). 4th ed. Eugene: University of Oregon Press; 2000

[99] Broeder MJ, Kopylova VA, Kamminga LM, Legler J. Zebrafish as a model to study the role of peroxisome proliferating-activated receptors in adipogenesis and obesity. PPAR Research. 2015;2015:1-11

[100] Forn-Cuni G, Varela M, Fernandez-Rodriguez CM, Figueras A, Novoa B. Liver immune responses to inflammatory stimuli in a diet-induced obesity model of zebrafish. Journal of Endocrinology. 2014;224:159-170

[101] Stankiewicz A, Mcgowan E, Yu L, Zhdanova I. Impaired sleep, circadian rhythms and neurogenesis in diet-induced premature aging. International Journal of Molecular Sciences. 2017;18:2243

[102] Behra M, Cousin X, Bertrand C, Vonesch J, Biellmann D, Chatonnet A, Strähle U. Acetylcholinesterase is required for neuronal and muscular development in the zebrafish embryo. Nature Neuroscience. 2002;5:111-118

[103] Schliebs R, Arendt T. The cholinergic system in aging and neuronal degeneration. Behavioural Brain Research. 2011;221:555-563

[104] Ding Y, Sun X, Huang W, Hoage T, Redfield M, Kushwaha S, et al. Haploinsufficiency of target of rapamycin attenuates cardiomyopathies in adult zebrafish. Circulation Research. 2011;109:658-669

[105] Hardy S, Legagneux V, Audic Y, Paillard L. Reverse genetics in eukaryotes. Biology of the Cell. 2010;102:561-580

[106] Gene Tools LLC. Morpholino Antisense Oligos [Internet]. Available from: http://www. gene-tools.com/morpholino_antisense_oligos [Accessed: 2018-01-29]

[107] Yuan S, Sun Z. Microinjection of mRNA and morpholino antisense oligonucleotides in zebrafish embryos. Journal of Visualized Experiments: JoVE. 2009;**27**:1113

[108] Rosen JN, Sweeney MF, Mably JD. Microinjection of zebrafish embryos to analyze gene function. Journal of Visualized Experiments: JoVE. 2009;**25**:1115

[109] Mimoto MS, Christian JL. Manipulation of gene function in *Xenopus laevis*. Vertebrate Embryogenesis: Embryological, Cellular, and Genetic Methods. 2011;**770**:55-75

[110] Ferguson DP, Dangott LJ, Lightfoot JT. Lessons learned from vivo-morpholinos: How to avoid vivo-morpholino toxicity. BioTechniques. 2014;**56**:251

[111] Ferguson DP, Schmitt EE, Lightfoot JT. Vivo-morpholinos induced transient knockdown of physical activity related proteins. PLoS One. 2013;**8**:e61472

[112] Moulton JD, Jiang S. Gene knockdowns in adult animals: PPMOs and vivo-morpholinos. Molecules. 2009;**14**:1304-1323

[113] Li YF, Morcos PA. Design and synthesis of dendritic molecular transporter that achieves efficient in vivo delivery of morpholino antisense oligo. Bioconjugate Chemistry. 2008;**19**:1464-1470

[114] Morcos PA, Li Y, Jiang S. Vivo-Morpholinos: A non-peptide transporter delivers Morpholinos into a wide array of mouse tissues. BioTechniques. 2008;**45**:613-614

[115] Gene Tools LLC. Vivo-Morpholinos [Internet]. Available from: http://www.gene-tools.com/vivomorpholinos [Accessed: 2018-01-29]

[116] Kizil C, Brand M. Cerebroventricular microinjection (CVMI) into adult zebrafish brain is an efficient misexpression method for forebrain ventricular cells. PLoS One. 2011;**6**:e27395

[117] Kizil C, Iltzsche A, Kaslin J, Brand M. Micromanipulation of gene expression in the adult zebrafish brain using cerebroventricular microinjection of morpholino oligonucleotides. Journal of Visualized Experiments: JoVE. 2013;**75**:e50415

[118] Nasevicius A, Ekker SC. Effective targeted gene 'knockdown' in zebrafish. Nature Genetics. 2000;**26**:216-220

[119] Corey DR, Abrams JM. Morpholino antisense oligonucleotides: Tools for investigating vertebrate development. Genome Biology. 2001;**2**:1015-1011

[120] Thummel R, Bai S, Sarras MP, Song P, McDermott J, Brewer J, et al. Inhibition of zebrafish fin regeneration using in vivo electroporation of morpholinos against fgfr1 and msxb. Developmental Dynamics. 2006;**235**:336-346

[121] Kizil C, Otto GW, Geisler R, Nüsslein-Volhard C, Antos CL. Simplet controls cell proliferation and gene transcription during zebrafish caudal fin regeneration. Developmental Biology. 2009;**325**:329-340

8

The Roles of Estrogen, Nitric Oxide, and Dopamine in the Generation of Hyperkinetic Motor Behaviors in Embryonic Zebrafish (*Danio rerio*)

Conor Snyder, Reid Wilkinson, Amber Woodard, Andrew Lewis, Dallas Wood, Easton Haslam, Tyler Hogge, Nicolette Huntley, Jackson Pierce, Kayla Ranger, Luca Melendez, Townsend Wilburn, Brian Kiel, Ty Krug, Kaitlin Morrison, Aaliayh Lyttle, Wade E. Bell and James E. Turner

Abstract

Both estrogen (E2) and nitric oxide (NO) have been shown to affect motor function, in part, through regulation of dopamine (DA) release, transporter function, and the elicitation of neuroprotection/neurodegeneration of healthy neurons, as well as in neurodegenerative conditions such as Parkinson's disease (PD). Currently, the "gold standard" treatment for PD is the use of levodopa (L-DOPA). However, patients who experience long-term L-DOPA and a monamine oxidase inhibitor (MAOI) treatment may develop unwanted side effects such as hyperkinesia which can be exacerbated by female Parkinsonian patients also on E2 replacement therapy. The current study was designed to determine whether embryonic zebrafish treated with either E2 or L-DOPA/MAOI develop a de novo-induced hyperkinetic movement disorder that relies on the NO pathway to elicit this hyperkinetic phenotype. Results from this study indicate that 5 days post-fertilization (dpf), fish treated with an L-DOPA + MAOI co-treatment or E2 elicited the development of a de novo hyperkinetic phenotype. In addition, the de novo L-DOPA + MAOI- and E2-induced hyperkinetic phenotypes are dependent on NO and E2 for its initiation and recovery. In conclusion, these findings point to the central role both NO and E2 play in the facilitation of de novo hyperkinesia.

Keywords: nitric oxide, estrogen, motor dysfunction, dopamine, L-DOPA, monoamine oxidase inhibitor, zebrafish

1. Introduction

Movement disorders are prominent symptoms of a number of neurodegenerative diseases such as Parkinson's disease (PD), Huntington's, and Alzheimer's disorders. For example, PD affects the motor system of the brain due to dopamine (DA) neurotransmitter deficiency. This disease is caused by the systematic degeneration of DA neurons in the basal ganglia of the brain [1]. Those with this movement disorder exhibit tremors, bradykinesia (hypokinesia), rigidity, balance and posture impairment, loss of automatic movements, and speech difficulties. PD affects millions across the world; the European Parkinson's Disease Association states that 6.3 million people have the neurodegenerative disorder globally [2]. Those who suffer with PD are without a cure and must resort to methods of PD treatment for relief. Currently, the "gold standard" treatment for PD is the use of levodopa (L-DOPA). A precursor to dopamine, L-DOPA is a small enough molecule to pass the blood-brain barrier and enter the basal ganglia where it is acted upon by DOPA decarboxylase to create an increase in dopamine levels. As DA neurons degenerate, an influx of dopamine from exogenous L-DOPA reverses the negative effects of PD [3]. In conjunction with L-DOPA, monoamine oxidase inhibitors (MAOI) are used to also increase dopamine levels as a co-treatment by inhibiting the DA-degrading enzyme monoamine oxidase. Thus, inhibiting monoamine oxidase in conjunction with L-DOPA treatment creates higher levels of DA in PD patients to help alleviate their symptoms. However, patients who experience long-term L-DOPA and MAOI treatment may develop unwanted side effects such as hyperkinesia, an increase in muscular activity that may be excessive or abnormal [4].

Previous studies have suggested that estrogen (E2) has neuroprotective effects in DA neurons and can regulate the synthesis of DA as a pro-dopaminergic agent [5]. In addition, studies show that DA neurons of the central nervous system have E2 receptors and the presence of the E2 synthesis enzyme aromatase [5]. It is clear that there is a connection between E2, the central nervous system, and movement disorders like PD. Indeed, premenopausal women are less likely to show PD symptoms with a majority of patients being male and over 60 [6]. Thus, there appears to be a sexual dimorphism between males and females when it comes to PD prevalence [6]. As a result of the hormonal differences, E2 is considered a neuroprotectant molecule, but there is no evidence for a similar role for testosterone [6]. Recently, this effect has been examined in female rats which have been treated with the 1-methyl-4-phenyl-1,2,3,6-tetrahydropyridine (MPTP) neurotoxin and have shown the ability to resist muscular activity loss compared to males [6]. In addition to being neuroprotective, there is also accumulating evidence that E2 may also cause detrimental effects such as hyperkinetic/chorea/dystonia symptoms in females on postmenopausal replacement therapy after hysterectomy [5]. There is also the recent case of a patient suffering from adult-onset Sydenham's chorea who discontinued E2 replacement therapy and months later these hyperkinetic/chorea symptoms were significantly diminished [7].

Part of the mechanism by which E2 may exert its influence in the nigrostrital (BG) of PD patients is through its documented influence on nitric oxide (NO) levels through its regulation of the expression of nitric oxide synthase (NOS) [8]. NO, a gas released by the actions of the NOS enzyme on L-arginine, acts as a signaling molecule with direct actions on existing metabolic pathways, as well as through genomic mechanisms [9, 10]. As a gas, NO can diffuse across cellular membranes without the aid of membrane-bound transport proteins or receptors. NO can interact directly with its end targets either in the cell in which it was synthesized or in surrounding cells. In turn, its actions are precisely controlled due to its very short half-life and restricted diffusion distance [11, 12]. At higher concentrations NO can act as a free radical in some situations or binds to superoxide anion (O_2^-), causing pathophysiological oxidative stress effects [13]. It is under these conditions that NO is thought to play a role in the genesis of such neurological diseases as PD [4]. On the other hand, NO at lower concentrations can act as a cellular protectant through prevention of apoptosis, excitotoxicity, neuronal depolarization, and regulation of the redox state in the mitochondria [14, 15]. In particular, NO has been implicated in the neuromodulation/neuroprotection of DA neurons in the nigrostrital (BG) pathway associated with either animal models of 1-methyl-4-phenyl-1,2,3,6-tetrahydropyridine (MPTP) or 6-hydroxydopamine (6OHDA) neurotoxicity that create PD-like symptoms or from PD patient clinical data. NO acting at the cellular level interacts with either its soluble guanylyl cyclase (sGC) receptor molecule to produce cyclic GMP (cGMP) which activates a cascade of cellular enzymes or causes S-nitrosylation of cysteine residues leading to protein conformational changes [16, 17]. These two pathways are referred to as either the NO-sGC-cGMP-dependent or NO-sGC-cGMP-independent pathways, respectively. In the BG, one of the four nitric oxide synthase (NOS) isoforms, neuronal (n)NOS, is believed to act through the NO-sGC-cGMP-dependent pathway which acts to modulate transcription factors, phosphodiesterases, ion-gated channels, or cGMP-dependent protein kinases (PKG), each of which continues to act physiologically in the nervous system [18]. In the BG, NO has been shown to affect DA release, influence transporter function, and elicit neuroprotection of DA neurons [19].

Zebrafish (*Danio rerio*) have been found to be an excellent model for studying motor disorders because they show similar neurological functions that humans possess and can easily demonstrate PD-like symptoms with damage to its basal ganglia-like structures [20]. In turn, a model where BG-like pathways are simpler and the DA neurons fewer in number and easier to visualize and access would be ideal for such studies. The embryonic zebrafish would appear to fit these criteria. The DA system has been well characterized in both embryonic development and in adults. The DA system in zebrafish, which is equivalent to the nigrostrital pathway in mammals, has been shown to ascend to the subpallium (striatum) from the basal diencephalon [21]. Also, zebrafish embryos and adults respond to the DA neurotoxins MPTP and 6OHDA, as well as to the DA receptor agonists/antagonists in much the same manner as in mammalian models of PD [22–24]. Indeed, there are an increasing number of studies which make a case for the use of zebrafish as a model for the study of movement disorders such as PD [20]. Earlier observations from our laboratory have established a zebrafish locomotor dysfunctional hypokinetic model linked to both E2 and NO deficiency [24]. In our most recent study, it was demonstrated that when NO synthases are inhibited in zebrafish, using nNOSI, a condition called "listless" occurs where the fish lack swimming abilities, are rigid, and have difficulty maintaining balance, similar to human symptoms of PD [25]. Also, co-treatment with either nNOSI or estrogen (E2), an upstream regulator of NO synthase, could rescue fish from the "listless"/PD-like phenotype caused by exposure to the neurotoxin 6-hydroxydopamine (6 OHDA) [26]. In turn, NO-deprived zebrafish were rescued from the "listless"/PD-like phenotype when co-treated with L-DOPA, a precursor to DA used routinely in PD therapy. Most significantly, NO involvement in the motor homeostasis of the embryonic zebrafish was shown to be expressed through the NO-sGC-cGMP-PKG-dependent pathway [26]. Therefore, initial evidence for NO's E2-linked role in locomotor activities in an embryonic zebrafish model was established.

It is the hypothesis of this study that when embryonic zebrafish are treated with either E2 or L-DOPA/MAOI that a de novo-induced hyperkinetic movement disorder phenotype will be generated. In conclusion, these results establish a rapid turnover zebrafish model for the study of the role of NO-E2-related DA actions in normal and hyperkinetic movement phenotypes.

2. Materials and methods

2.1. Fish preparations

The compound *roy;nacre* double-homozygous mutant zebrafish, named *casper*, were used in this study. Casper shows the effect of combined melanocyte and iridophore loss in which the body of the embryonic and adult fish is largely transparent due to loss of light absorption and reflection [27]. These transgenic fish were obtained from Carolina Biological Supply. All fish were maintained in a basic embryonic rearing solution (ERS) consisting of NaCl, $CaCl_2$, KCl, and $MgSO_4$. These necessary ions were dissolved in deionized water containing a 0.05% methylene blue solution, which serves as an antimicrobial agent. All solutions were changed every 24 h, and embryos were incubated at 28°C. All reagents were obtained from Sigma-Aldrich, and solutions were made daily before use unless noted otherwise. Fish treated at 4–6 days post-fertilization (dpf) were allowed to hatch on their own prior to treatment. All procedures were in accordance with NIH guidance for the care and treatment of animals.

2.2. Reagent preparations

2.2.1. E2-related reagents

All E2-related reagents for treating zebrafish have been previously tested in a dose-response paradigm to insure optimal results and proper survival [26]. Based on previous studies, E2 (17β-estradiol, Sigma) used at 1 and 5 μM, and initially solubilized in a 100% ethanol stock solution diluted down to the base treatment solution with ERS, ensuring that the ethanol concentration in the final solution was equal to or lower than 0.5%. The control group consisted of ERS salt solution plus 0.5% ethanol. The reagent 4-androstene-3,17-dione (4-OH-A, MW-286.4, Sigma) was used as an aromatase inhibitor (AI) to block the production of E2 from androgens [24, 25, 28]. It was used at 50 μM and made from a 100% ethanol stock solution diluted down to the base treatment solution with ERS, ensuring that the ethanol concentration in the final solution was equal to or lower than 0.5%.

2.2.2. NO-related reagents

All NO-related reagents for treating zebrafish have been previously tested in a dose-response paradigm to insure optimal results and proper survival. Based on literature review, baseline target concentrations were identified. Proadifen hydrochloride (Sigma) was used as a selective nNOS inhibitor (nNOSI). With ERS as the diluent, fish were tested at 10, 30, and 50 μM. The 50 μM concentration provided optimal results in its ability to create the hypokinetic (listless) condition, and this dose was used throughout the current study.

Diethylenetriamine/nitric oxide adduct (DETA-NO, Sigma) was used to provide a slow extended release of exogenous NO as a co-treatment with some of the inhibitors used in the experiments in an effort to show that NO inhibition-mediated symptoms exhibited by fish can be rescued. It was dissolved into ERS resulting in working concentrations of 400–50 μM with 50 μM providing the best results.

1H-[1,2,4]Oxadiazolo[4,3-a]quinoxalin-1-one (ODQ, Sigma) was used as a soluble guanylyl cyclase (sGC) inhibitor which compromises the NO-cGMP-dependent pathway by reducing cGMP production. It was dissolved into a 0.1% DMSO solution and then diluted with ERS to a working concentration of 30 μM for application. In addition, DTT (dithiothreitol, Sigma) was used as an inhibitor of the NO-cGMP-independent pathway which prevents S-nitrosylation events at a concentration of 100 μM.

2.2.3. DA-related reagents

The L-DOPA DA precursor L-DOPA ethyl ester (L-3, 4-dihydroxyphenylalanine methyl ester, Sigma) are used at concentrations up to 10 mM, which is the limit of its solubility in the ERS control solution. L-DOPA is acted upon by DOPA decarboxylase to be converted into DA. It was used to elevate the neurotransmitter in deficient fish starting at ranges prescribed previously for zebrafish embryos [29, 30]. The optimal dose was 10 mM and used throughout the current study.

Monoamine oxidase inhibitor (MAOI) is an agent used to manipulate the zebrafish DA neurons by preventing DA degradation at the synapse. The MAOI, L-deprenyl (Sigma), was used at a concentration of 50 μM to elicit hyperdyskinetic behavior in a co-treatment paradigm with L-DOPA.

A DA receptor antagonist (haloperidol, Sigma) was used at a concentration of 1–50 μM as prescribed for zebrafish embryos [31] and 1 μM was found to be optimal.

2.3. Hyperkinesia phenotype protocols

Fish at 5 dpf were co-treated with L-DOPA + MAOI for up to 48 h or E2 alone for 3–6 h to induce a hyperkinetic state. Using this protocol, additional experiments were also designed to determine whether either L-DOPA or MAOI alone could cause the hyperkinetic phenotype. Specifically, fish were treated with either L-DOPA, MAOI, or a L-DOPA + MAOI co-treatment along with the ERS controls. Next, studies were designed to determine if the hyperkinetic phenotype could be modified changing NO levels in the L-DOPA + MAOI-treated fish. Specifically, the co-treatment (L-DOPA + MAOI) was compared to the L-DOPA + MAOI + nNOSI tri-treatments along with their respective controls. Next, experiments were designed to test recovery of 5 dpf fish after a 40-h treatment with L-DOPA + MAOI which was followed by either ERS, nNOSI, or DETA-NO post-treatment washouts. The third set of experiments looked at the role of E2 in the generation of the hyperkinetic state. Specifically, fish were treated with either E2, at various concentrations, L-DOPA + MAOI, and L-DOPA + MAOI + AI. Similar co-treatment studies were carried out with E2 according to the following protocols: E2 + haloperidol and E2 + nNOSI.

2.4. Data collection

For visual analysis, fish were characterized using a dissecting microscope, as expressing the hyperkinetic dyskinesia phenotype when their swimming behaviors became significantly different from ERS controls. Specifically, the 'hyperkinetic dyskinesia' phenotype was identified as showing rapid, erratic, and brief spurts of swimming movements and was either calculated as a percent of the treated group or by video capture analysis using a Nikon SMZ1500 microscope to measure the number of spontaneously initiated swimming movements per minute. Also, fish were timed (seconds) as to the duration of their startle/escape response to being touched by a probe on the tail region. The percent survival under the various experimental conditions was also determined for both the hyperkinetic treatment conditions.

2.5. Data analysis

Data were analyzed for significant differences either by a z-test for two-population proportions or for multiple proportions using chi-square contingency table test, followed by a Marascuilo's post-hoc analysis. In addition, for timed video capture movements and startle/escape responses, statistical analysis by using either a two-tailed t-test or an analysis of variance (ANOVA) one-way paired t-test or ANOVA Single Factor tests. For the ANOVA analysis a Tukey post-hoc method was also run to determine significant differences between the various treatment groups. Sample sizes for all separately treated fish were $n = 30$ and all experiments were repeated in triplicate.

3. Results

3.1. L-DOPA + MAOI co-treatment cause the development of a de novo hyperkinetic phenotype in 5 dpf fish

Figure 1A shows the percentage of zebrafish that demonstrated a hyperkinetic phenotype when co-treated with L-DOPA + MAOI over 40 h of treatment compared with ERS controls. These data show that a significant portion of a population exhibited hyperkinesia after 24 h (55%) in the co-treatment and rises to 90% after 40 h ($p < 0.01$) compared to 0% for the ERS controls. This provides evidence that the co-treatment is an effective combination for inducing the hyperkinetic phenotype compared to ERS controls that demonstrated no hyperkinesia. Specifically, the hyperkinetic fish demonstrated spontaneous swift, erratic, and chorea/catatonic excitement-like movements when compared to controls.

Figure 1B demonstrates that co-treated fish remain stable for the duration of the treatment paradigm with no significant deaths when compared to ERS controls ($p > 0.05$).

Figure 2 shows photomicrographs from video capture of zebrafish fin movements under various treatment conditions. Note that control fish exhibited synchronous and symmetrical adduction (**Figure 2A**) and abduction (**Figure 2B**) of fin positions during movement or at rest. In contrast, the L-DOPA + MAOI co-treated fish show asymmetric and asynchronous adduction and abduction in their pectoral movements (**Figure 2C**). Behaviorally, these fish exhibit a lack of control of swimming movements and chorea/catatonic excitement-like symptoms.

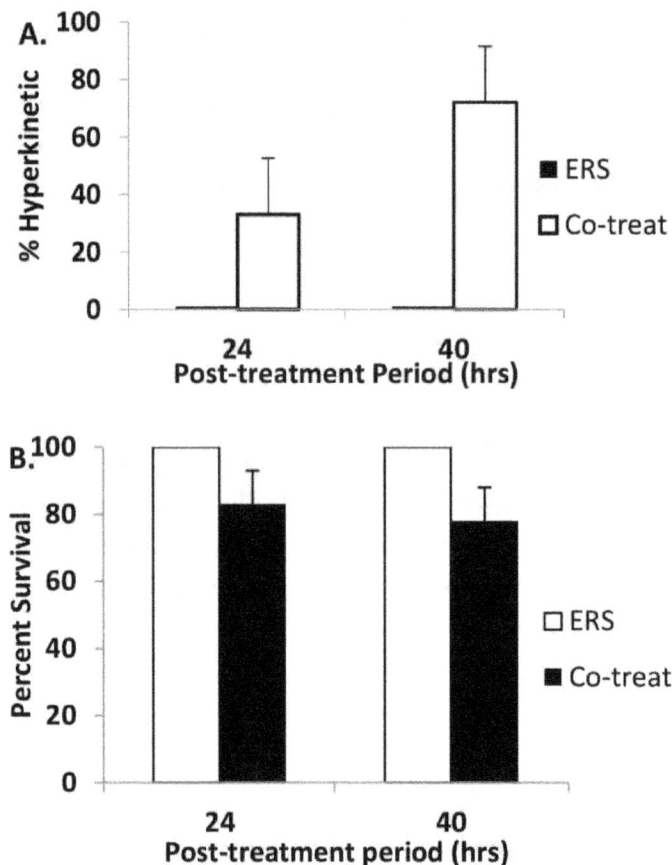

Figure 1. The effects of an L-DOPA + MAOI co-treatment on generating a hyperkinetic motor behavior phenotype in 5 dpf zebrafish evaluated at 24 and 40 h post-treatment. (A) The percentage of hyperkinetic zebrafish induced by co-treatment with L-DOPA + MAOI compared to ERS controls over 24- and 40-h treatment periods shows a significant increase (*$p < 0.01$) in the appearance of the hyperkinetic phenotype when compared to ERS controls. (B) Shows that there was no significant difference between the ERS and co-treatment mortality rates at both 24 and 48 h post-treatment ($p > 0.05$). Bars = ±SD.

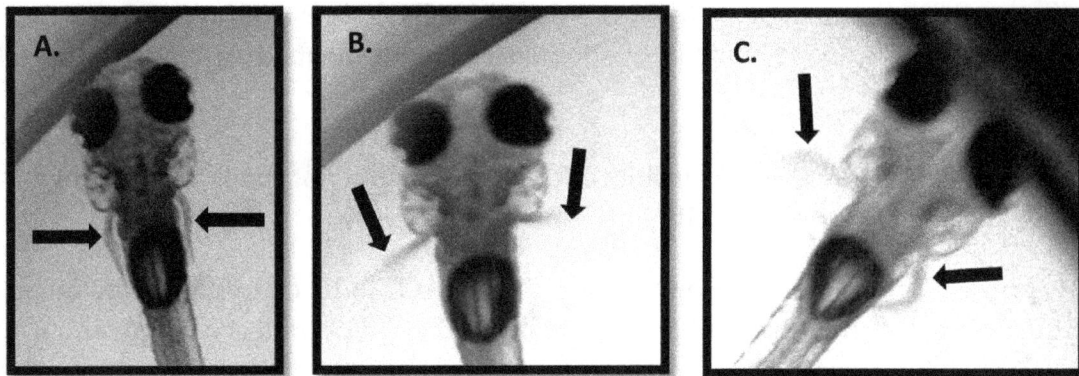

Figure 2. Video capture photomicrographs of 5 dpf fish fin movements 40 h post-treatment. (A and B) Pictures represent ERS control zebrafish pectoral fin synchronous movements in the adduction (A) and abduction (B) states (arrows). (C) Asymmetric pectoral fin movements in response to L-DOPA + MAOI co-treatment. Note that while the left fin is abducted, the right one is adducted (arrows).

3.2. MAOI is more effective than L-DOPA in eliciting the de novo hyperkinetic phenotype in 5 dpf fish

Figure 3 shows the frequency of initiated movements among ERS control, L-DOPA, MAOI, and co-treatment (L-DOPA + MAOI) fish. This experiment tested which of the two DA-related reagents in the co-treatment was more responsible for generating the hyperkinetic phenotype. These data show that MAOI is the primary facilitator of de novo hyperkinesia in the co-treatment when compared to L-DOPA ($p < 0.01$). Specifically, control and L-DOPA frequency of spontaneous movements were not significantly different at 40 h of treatment (41 ± 5 compared to 37 ± 12 times/min, respectively, $p > 0.05$). Similarly, L-DOPA + MAOI and MAOI-only treatments were not significantly different (74 ± 15 compared to 73 ± 12 times/min, $p > 0.05$). However, MAOI treatment was significantly different from L-DOPA treatment (37 ± 12 compared to 73 ± 12

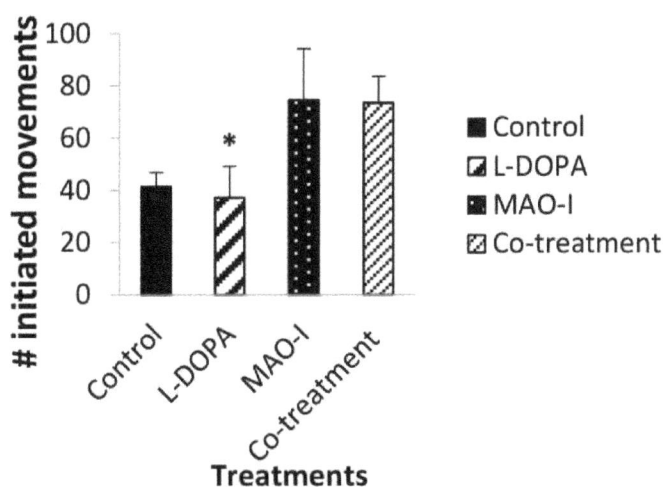

Figure 3. The number of spontaneous movements that 5 dpf zebrafish initiated over a 1-min duration after 40 h under various treatment conditions. Note that ERS controls initiated movements at approximately 41 times/min, L-DOPA treated zebrafish at 37 times/min, MAOI at 74 times/min, and co-treated L-DOPA/MAOI at 73 times/min. Bars = ±SD. *$p < 0.01$.

times/min, $p < 0.01$). Specifically, both MAOI and the L-DOPA + MAOI co-treatment initiated by approximately twofold the number of spontaneous movements than that of either ERS or L-DOPA alone.

3.3. The de novo L-DOPA + MAOI-induced hyperkinetic phenotype is dependent on NO, E2, and the DA system for its initiation and recovery

Figure 4 shows the effect of nNOSI on the L-DOPA + MAOI-induced hyperkinetic phenotype. During the first 48 h of treatment, ERS controls showed no hyperkinesia; however, L-DOPA + MAOI co-treatment demonstrated 94% hyperkinesia. Note that **Figure 4A** shows that the L-DOPA + MAOI + nNOSI tri-treatment significantly reduced the co-treatment-induced hyperkinesia after 48 h of treatment (43% vs. 94%, respectively, $p < 0.01$). **Figure 4B** shows the duration of zebrafish swimming, post-tail probe, comparing ERS control, L-DOPA + MAOI

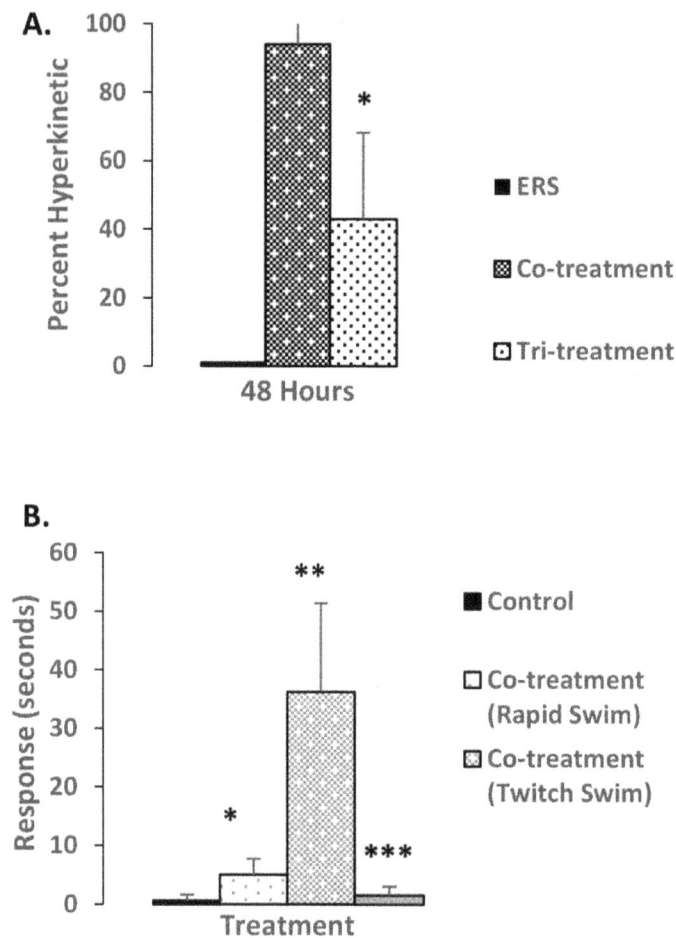

Figure 4. The effect of nNOSI on the L-DOPA + MAOI-induced hyperkinetic phenotype at 48 h post-treatment. (A) The tri-treatment (L-DOPA + MAOI + nNOSI caused a significant reduction (*$p < 0.01$) in the percent of fish demonstrating the tail probe-induced hyperkinetic phenotype when compared to the co-treatment (L-DOPA + MAOI). (B) The duration of zebrafish swimming, post-tail probe, comparing ERS control, L-DOPA + MAOI co-treatment, and L-DOPA + MAOI + nNOSI tri-treatments. Note that the tri-treatment significantly decreased the swim duration (***$p < 0.05$) when compared to the co-treatment values. Also note that two distinct swimming behaviors, rapid followed by twitch-swim, were observed only in co-treated fish in response to probe stimulation and were significantly different from ERS control values and from each other (*$p < 0.05$ rapid swim, and **$p < 0.05$. Twitch-swim). Bars = ±SD.

co-treatment, and L-DOPA + MAOI + nNOSI tri-treatments. Note that co-treatment escape swimming behaviors demonstrate a rapid swim followed by much longer twitch swim duration. Specifically, the co-treatment rapid swim phenotype had an average swim duration of 5.1 ± 2.5 s followed by a twitch swim duration of 36.2 ± 12.5 s. In contrast, the L-DOPA + MAOI + nNOSI tri-treatment was significantly reduced the swim duration to 1.5 ± 0.2 s when compared to the co-treatment values ($p < 0.01$).

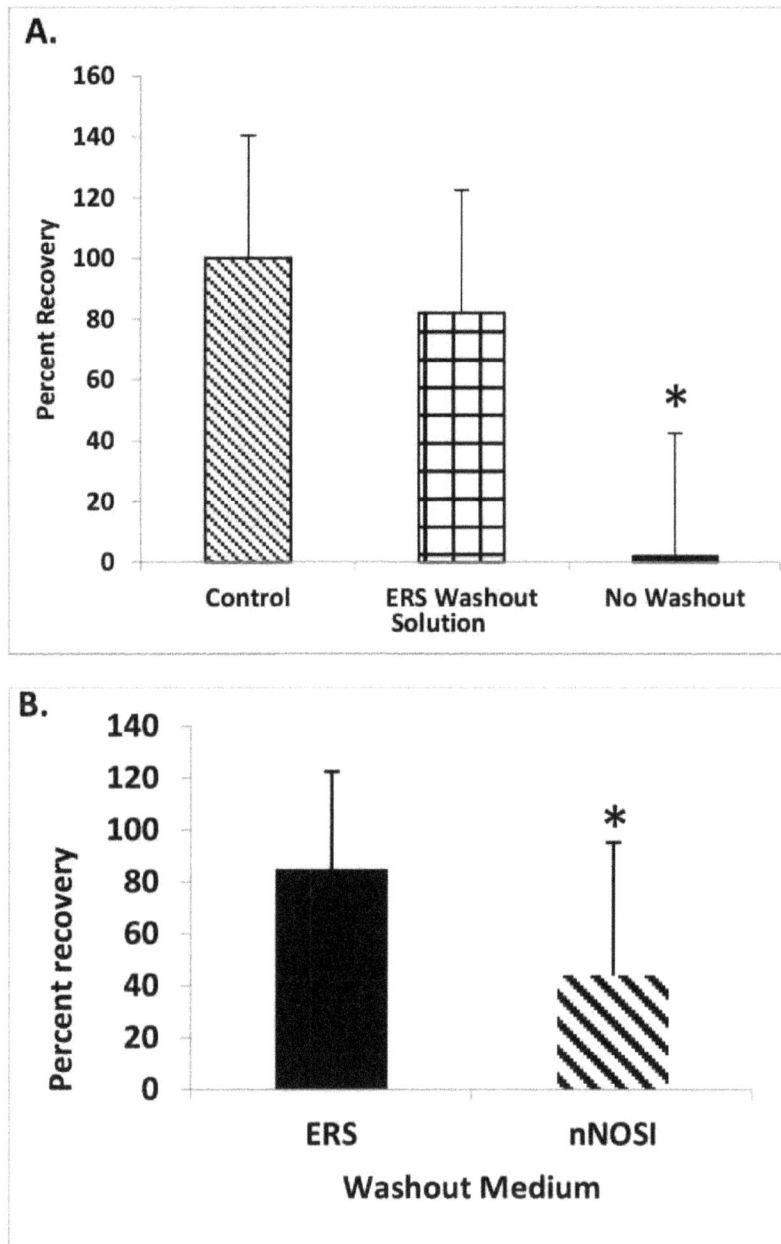

Figure 5. Demonstration of the percentages of 5 dpf fish that recovered after 24 hours of post-treatment (washout) with various treatments from the hyperkinetic state initially induced by the co-treatment of L-DOPA + MAOI. (A) The washout with ERS showed that approximately 80% of the fish recovered when compared to 0% of the non-washed out fish. At this point the ERS washout fish were not significantly different from those of controls (*$p > 0.05$). (B) Recovery rate of hyperkinetic fish in response to ERS, and nNOSI post-treatments (washout). Note that nNOSI post-treatment significantly inhibits the recovery response when compared to either ERS or DETA-NO post-treatments (*$p < 0.01$). Bars = $\pm$SD.

Figure 5 depicts experiments testing recovery of 5 dpf fish after a 40 hours treatment with L-DOPA + MAOI to induce the hyperkinetic phenotype. Data were collected after a 24 hours post-treatment washout with either ERS or nNOSI. **Figure 5A** shows that fish treated continually with ERS demonstrates normal (non-hyperkinetic) swimming behaviors and the fish that were not washed out with ERS (just kept in the co-treatment) had a 0% recovery rate. However, the fish that were washed out with ERS solution after the initial co-treatment had approximately an 80% recovery back to normal swimming patterns. **Figure 5B** shows that post-treatment washout with nNOSI post-treatment (less than 20%) washouts showed significantly less recovery ($p < 0.01$).

Figure 6 shows what happens to swimming durations when at 5 dpf, zebrafish were treated with different concentrations of E2. At 6 h post-treatment, fish were lightly touched with a probe and their escape response timed (s—seconds) until they stopped. When the ERS control fish were stimulated they swam for 0.5 ± 0.3 s. Fish treated with 1 µM E2, responded by swimming 1.2 ± 0.5 s, which was not significantly ($p > 0.05$) different when compared to the ERS controls. However, fish treated with a 5 µM E2 dosage, swam at durations significantly longer (4.7 ± 3.9 s, $p < 0.05$) when compared to both the ERS and standard 1 µM dosage of E2. Specifically, 100% of the higher dosage 5 µM E2 fish exhibited significant hyperkinetic activity. Conversely, fish treated with 10 µM E2 were not able to survive the treatment.

When exposed to various treatments with the reagents AI, and L-DOPA + MAOI or L-DOPA + MAOI + AI, 5 dpf fish exhibited several different swimming phenotypes (**Figure 7**). Specifically, fish treated with AI were 67% listlessness, a significantly higher proportion ($p < 0.05$) when compared to the two other group's swimming characteristics. Most significantly, hyperkinesia

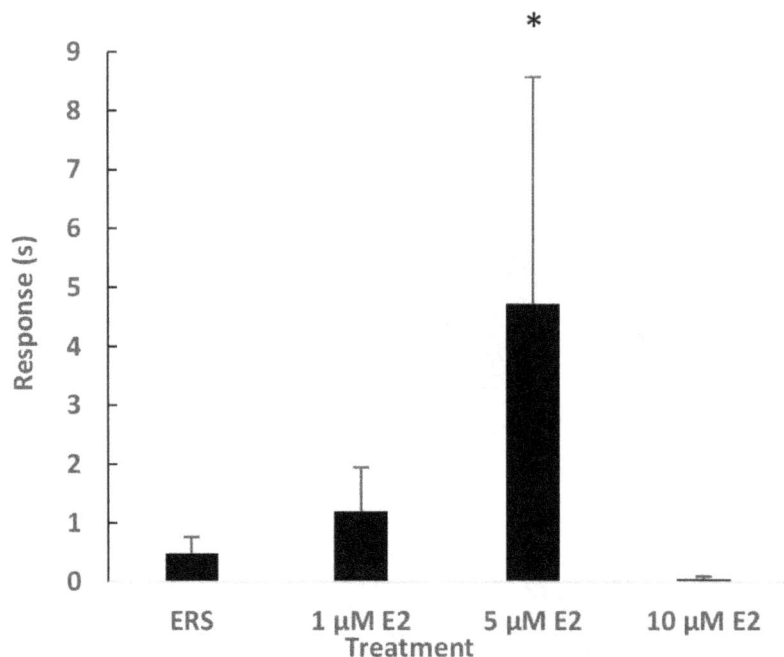

Figure 6. Zebrafish at 5 dpf are treated with different concentrations of E2 and at 6 h post-treatment, fish were lightly touched with a probe and their escape response timed (s—seconds) until they stopped swimming. Note that a 5 µM E2 concentration caused a significant increase in hyperkinetic swimming activity when compared to a 1 µM E2 dose (*$p < 0.05$). A 10 µM dose of E2 was found to be toxic. Bars = ±SD.

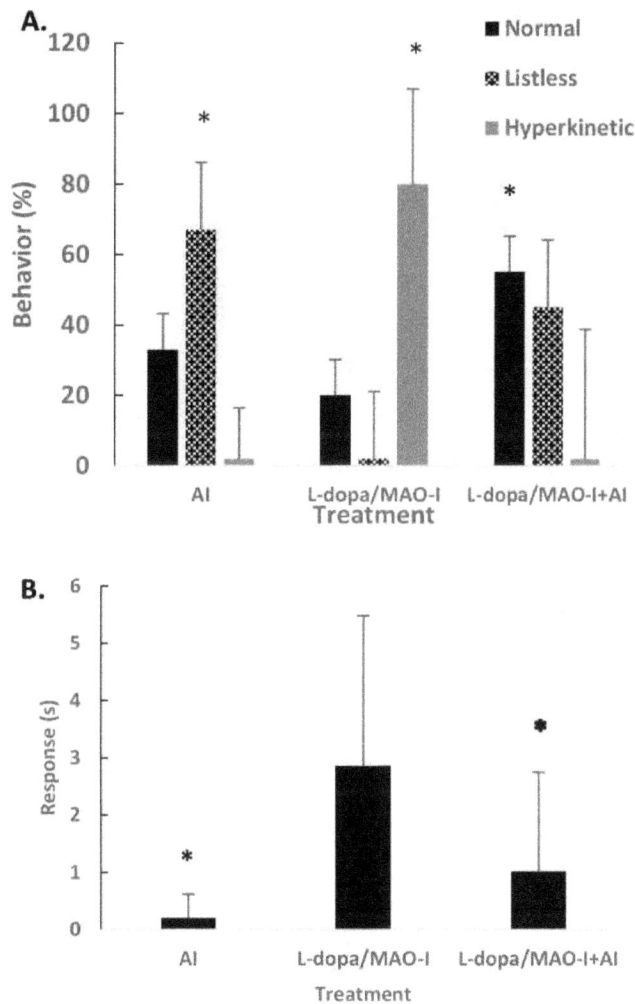

Figure 7. Effects of AI treatment on L-DOPA + MAOI-induced hyperkinetic activity in 5 dpf fish. Note that the addition of AI to the tri-treatment (L-DOPA + MAOI + AI) significantly reduced (*$p < 0.01$) the hyperkinetic probe-induced swim time (s—seconds) when compared to that of the co-treatment (L-DOPA + MAOI). Bars = ±SD.

was seen in approximately 80% of the co-treated (L-DOPA + MAOI) fish, a significant percentage ($p < 0.05$) when compared to negligible percentages in both the AI fish and the tri-treated (or L-DOPA + MAOI + AI) fish (**Figure 7A**). Normal swimming behavior was the dominant phenotype in the tri-treated fish with a significant ($p < 0.05$) proportion of fish exhibiting this behavior. **Figure 7B** shows that AI treated fish exhibited a significantly reduced swimming duration ($p < 0.05$) when compared to both the co-treatment and tri-treatments. The addition of AI to the L-DOPA + MAOI co-treatment significantly reduced the response time of fish exhibiting the hyperkinetic phenotype. Specifically, co-treated fish (L-DOPA + MAOI) swam at 2.9 ± 0.4 s when probed which is much quicker when compared to the AI (0.2 ± 0.4 s) and tri-treated fish (1.0 ± 1.8 s).

Figure 8A shows the effect of HA on E2-induced hyperkinesia. Specifically, the addition of HA to E2 in a co-treatment paradigm (E2 + HA) significantly reduced ($p < 0.05$) the hyperkinetic probe-induced swim time by fourfold from 4.2 ± 3.9 s to 0.5 ± 0.3 s. In addition, the co-treatment values were very similar to those of the ERS (0.4 ± 0.2 s) and HA ((0.4 ± 0.2 s)

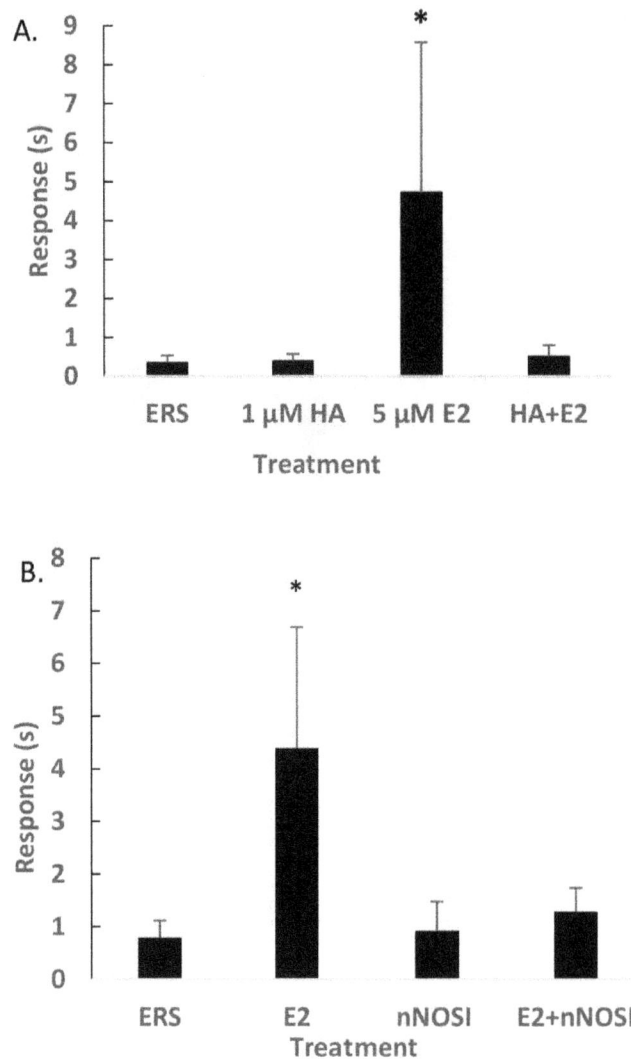

Figure 8. The effect of HA and nNOSI on E2-induced hyperkinesia. (A) Note that the addition of HA to E2 as a co-treatment significantly reduced (*$p < 0.05$) the hyperkinetic probe-induced swim time (s—seconds) by fourfold when compared to E2 values. Also, the co-treatment reduced the swim time values to approximately those of the ERS and HA controls. (B) Note that the addition of nNOSI to E2 as a co-treatment significantly reduced (*$p < 0.01$) the hyperkinetic probe-induced swim time (s—seconds) by threefold when compared to E2 values. Also, the co-treatment swim time values were not significantly different ($p > 0.05$) from those of the ERS and nNOSI controls. Bars = ±SD.

control values. **Figure 8B** shows the effect of nNOSI on E2-induced hyperkinesia. Specifically, the addition of nNOSI to E2 in a co-treatment paradigm (E2 + nNOSI) significantly reduced ($p < 0.01$) the hyperkinetic probe-induced swim time by three-fold from 4.39 ± 2.3 s to 1.3 ± 0.5 s. In addition, the co-treatment values were not significantly different ($p > 0.05$) to those of the ERS (0.8 ± 0.3 s) and nNOSI (0.9 ± 0.6 s) control values.

4. Discussion

The goal of this study was to explore the hypothesis that the co-treatment of L-DOPA + MAOI, and E2 by itself will produce a zebrafish model of de novo hyperkinesia which are both dependent on the NO pathway for its expression. Also, the current study explored the possibility of using nNOSI as a modulating agent to reduce the de novo hyperkinetic dyskinesia phenotype in the zebrafish.

Data from the current study shows that 5 dpf zebrafish exhibited hyperkinesia as early as 24 h after treatment with an L-DOPA + MAOI co-treatment. Specifically, the hyperkinetic fish demonstrated spontaneous swift, erratic, and chorea/catatonic excitement-like movements, as well as, a significant increase in the number of spontaneous movements when compared to controls. This is the first report of de novo L-DOPA + MAOI-induced hyperkinesia in embry-onic zebrafish. However, L-DOPA has been shown in older zebrafish larvae to facilitate recov-ery of swimming speed after treatment with the antipsychotic fluphenazine [29]. In addition, data from the current study also show that MAOI is the primary facilitator of hyperkinesia in the co-treatment when compared to L-DOPA. Specifically, both MAOI and the L-DOPA + MAOI co-treatment initiated by approximately twofold the number of spontaneous move-ments than that of either ERS or L-DOPA alone. In an interesting corollary to this finding, it was shown in an earlier study that L-DOPA administered to zebrafish reduced the number of neurons in its nigrostriatal-like pathway which was partially rescued by monoamine oxidase inhibition [32]. This study was focused on the possibility that L-DOPA contains a neurotoxic product that may cause oxidative stress to DA neurons. We saw none of these symptoms in our study perhaps due to the fact that our findings were collected over a matter of 1–2 days duration which was not long enough to see these potential side effects. On the other hand, the fact that monoamine oxidase inhibition increased fish motor activity by a post-treatment paradigm is in support of our findings [32].

The current study also reported that the de novo L-DOPA + MAOI-induced hyperkinetic phenotype is dependent on NO for its initiation and recovery. Specifically, the L-DOPA + MAOI + nNOSI tri-treatment significantly reduced the L-DOPA + MAOI co-treatment-induced hyperkinesia. Similar results in earlier studies have shown that in hemiparkinsonian rats nNOSI improves L-DOPA-induced dyskinesia [4]. The findings are also in line with earlier suggestions of the possibility that nNOSI could be used as a therapeutic agent to reduce the dyskinetic side effects of long-term L-DOPA therapy [33]. In turn, current post-treatment studies demonstrated that NO accelerates recovery from the L-DOPA + MAOI-induced hyperkinetic phenotype when compared to ERS controls. In contrast, nNOSI post-treatment significantly reduced the rate of recovery from the hyperkinetic phenotype. These findings are most likely explained by the documented effects of NO on DA dynamics. Specifically, in the BG, NO has been shown to affect DA release, influence transporter function, and elicit neuroprotection of DA neurons [19].

In the present study, it was also determined that E2 can cause a de novo hyperkinetic phenotype in zebrafish. Specifically, a 3–6 h treatment with E2 elicited a tenfold increase in fish swim duration when compared with that of ERS controls. E2 was also found to significantly affect the L-DOPA + MAOI co-treatment-induced de novo hyperkinetic phenotype. Specifically, the addition of AI to the L-DOPA + MAOI co-treatment significantly reduced the response time of fish exhibiting the hyperkinetic phenotype returning them back to control levels. These data suggest that E2 is linked to the DA system regulating motor activity in the embryonic zebrafish. This finding was further validated in this study by results showing that the DA receptor antagonist, haloperidol (HA), significantly diminished the E2-induced de novo hyperkinetic activity. Specifically, a co-treatment of E2 + HA significantly reduced by fourfold the hyperkinetic phenotype when compared to just an E2 treatment. This evi-dence leads to the conclusion that the E2-induced hyperkinetic phenotype acts through the DA D1/D2 receptor system. This conclusion is further substantiated by an earlier study that showed that HA significantly reduced the level of larval zebrafish locomotor activity along a similar time line [31]. Furthermore, the effects of E2 on stimulating/regulating DA levels and thus motor activity have been well documented in other animal models. Specifically, it has been shown that E2 influences DA dynamics in the nigrostrital pathway that is crucial for normal motor function and is the site of PD pathology [5]. In this system, similar to NO, E2 affects the synthesis, release and turnover of DA, as well as DA transporter and recep-tor expression [5]. E2 derivatives have also been shown to cause hyperactivity in animal models. Specifically, the addition of bisphenol A, a xenoestrogen exhibiting E2-mimicking hormone-like properties, was shown to cause hyperactivity in newborn mice, adult male rats, and larval zebrafish [34–36]. However, the present study reports for the first time a rapid de novo E2-induced hyperkinetic response over just a 3–6 h duration in the embryonic zebrafish. In addition, the current de novo E2-induced hyperkinetic zebrafish model appears to correlate with accumulating evidence that E2 may also cause detrimental effects such as hyperkinetic/chorea/dystonia symptoms in female patients either through postmenopausal replacement therapy or through E2 replacement therapy after hysterectomy [5]. There is also the recent case of a patient suffering from adult-onset Sydenham's chorea who discontinued E2 replacement therapy and months later these hyperkinetic/chorea symptoms were signifi-cantly diminished [7].

5. Conclusions

The current study was designed to determine whether embryonic zebrafish treated with either E2 or L-DOPA/MAOI would develop a de novo-induced hyperkinetic movement disor-der and that they rely on the NO pathway to elicit this hyperkinetic phenotype. Results from this study indicate that 5 dpf fish treated with an L-DOPA + MAOI co-treatment or E2 elicited the development of a de novo hyperkinetic phenotype. In addition, the de novo L-DOPA + MAOI- and E2-induced hyperkinetic phenotypes are dependent on NO and E2 for its initia-tion and recovery. In conclusion, these findings point to the central role that both NO and E2 play in the facilitation of de novo hyperkinesia. In turn, the actions of both E2 and L-DOPA + MAOI in the induction of the hyperkinetic phenotype is dependent on the NO pathway and acts through the DA system. Most significantly, nNOSI has the capacity in this model to modulate the de novo hyperkinetic phenotype which suggests the possibility that it may be further tested for its therapeutic value in patients suffering from long-term L-DOPA-induced dyskinetic side effects.

Acknowledgements

This research was supported from grant funding from the Reid '41 Institute Professorship in the Arts and Sciences (awarded to JET), the VMI Department of Biology, and the VMI Center for Undergraduate Research.

Conflicts of interest

The authors have no conflicts of interest.

Author details

Conor Snyder, Reid Wilkinson, Amber Woodard, Andrew Lewis, Dallas Wood, Easton Haslam, Tyler Hogge, Nicolette Huntley, Jackson Pierce, Kayla Ranger, Luca Melendez, Townsend Wilburn, Brian Kiel, Ty Krug, Kaitlin Morrison, Aaliayh Lyttle, Wade E. Bell and James E. Turner*

*Address all correspondence to: turnerje@vmi.edu

Department of Biology, Center for Molecular, Cellular, and Biological Chemistry, Virginia Military Institute, Lexington, VA, USA

References

[1] Stephenson-Jones M, Ericsson J, Robertson B, Grillner SJ. Evolution of the basal ganglia: Dual-output pathways conserved throughout vertebrate phylogeny. Journal of Comparative Neurology. 2012;**520**:2957-2973

[2] GBD 2013 Mortality and Causes of Death Collaborators. Global, regional, and national age-sex specific all-cause and cause-specific mortality for 240 causes of death, 1990-2013: A systematic analysis for the Global Burden of Disease Study 2013. Lancet. 2015; **385**:117-171

[3] Mazzucchi S, Frosini D, Bonuccelli U, Ceravolo R. Current treatment and future prospects of dopa-induced dyskinesias. Drugs Today (Barc). 2015;**51**:315-329

[4] Padovan-Neto FE, Echeverry MB, Chiavegatto S, Del-Bel E. Nitric oxide synthase inhibitor improves de novo and long-term l-DOPA-induced dyskinesia in Hemiparkinsonian rats. Frontiers in Systems Neuroscience. 2011;**5**:40. DOI: 10.3389/fnsys.2011.00040. eCollection 2011

[5] Cersosimo MG, Benarroch EE. Estrogen actions in the nervous system: Complexity and clinical implications. Neurology. 2015;**85**:263-273

[6] Smith KM, Dahodwala N. Sex differences in Parkinson's disease and other movement disorders. Experimental Neurology. 2014;**259**:44-56

[7] Delaruelle Z, Honore P-J, Santens F. Adult-onset Sydenham's chorea or drug-induced movement disorder? A case report. Acta Neurologica Belgica. 2016;**116**:399-400

[8] Chambliss KL, Shaul PW. Rapid activation of endothelial NO synthase by estrogen: Evidence for a steroid receptor fast-action complex (SRFC) in caveolae. Steroids. 2002; **67**:413-419

[9] Lima B, Forrester MT, Hess DT, Stamler JS. S-nitrosylation in cardiovascular signaling. Journal of the American Heart Association. 2010;**106**:633-646

[10] Pelster B, Grillitsch S, Schwerte T. NO as a mediator during the early development of the cardiovascular system in the zebrafish. Comparative Biochemistry and Physiology. Part A, Molecular & Integrative Physiology. 2005;**142**:215-220

[11] Bradley S, Tossell K, Lockley R, McDearmid JR. Nitric oxide synthase regulates morphogenesis of zebrafish spinal cord motor neurons. The Journal of Neuroscience. 2010; **30**:16818-16831

[12] Hammond J, Balligand JL. Nitric oxide synthase and cyclic GMP signaling in cardiac myocytes: From contractility to remodeling. Journal of Molecular and Cellular Cardiology. 2011;**52**:330-340

[13] Forstermann U, Sessa WC. Nitric oxide synthases: Regulation and function. European Heart Journal. 2012;**33**:829-837

[14] Karaçay B, Bonthius DJ. The neuronal nitric oxide synthase (nNOS) gene and neuroprotection against alcohol toxicity. Cellular and Molecular Neurobiology. 2015;**35**:449-446

[15] Kurauchi Y, Hisatsune A, Isohama Y, Sawa T, Akaike T, Katsuk H. Nitric oxide/soluble guanylyl cyclase signaling mediates depolarization-induced protection of rat mesencephalic dopaminergic neurons from MPP+ cytotoxicity. Neuroscience. 2013;**231**:206-215

[16] Gao Y. The multiple actions of NO. Pflügers Archiv: European Journal of Physiology. 2010;**459**:829-839

[17] Tota B, Amelio D, Pelligrino D, Ip YK, Cerra MC. NO modulation of myocardial performance in fish hearts. Comparative Biochemistry and Physiology. 2005;**142**:164-177

[18] Derbyshire ER, Marlett MA. Structure and regulation of soluble guanylyl cyclase. Annual Review of Biochemistry. 2012;**81**:533-559

[19] Lorenc-Koci E, Czarnecka A. Role of nitric oxide in the regulation of motor function. An overview of behavioral, biochemical and histological studies in animal models. Pharmacological Reports. 2013;**65**:1043-1055

[20] Flinn L, Bretaud S, Lo C, Ingham PW, Bandmann O. Zebrafish as a new animal model for movement disorders. Journal of Neurochemistry. 2008;**106**:1991-1997

[21] Rink E, Wullimann MF. The teleostean (zebrafish) dopaminergic system ascending to the subpallium (striatum) is located in the basal diencephalon (posterior tuberculum). Brain Research. 2001;**889**:316-330

[22] McKinley ET, Baranowski TC, Blavo DO, Cato C, Doan TN, Rubinstein AL. Neuro-protection of MPTP-induced toxicity in zebrafish dopaminergic neurons. Brain Research. Molecular Brain Research. 2005;**141**:128-137

[23] Parng C, Roy NM, Ton C, Lin Y, McGrath P. Neurotoxicity assessment using zebrafish. Journal of Pharmacological and Toxicological Methods. 2007;**55**:103-112

[24] Nelson B, Henriet RP, Holt AW, Bopp KC, Houser AP, Allgood OE, Turner E. The role of estrogen on the developmental appearance of sensory-motor behaviors in the zebrafish (*Danio rerio*): The characterization of the "listless" mode. Brain Research. 2008;**1222**: 118-128

[25] Allgood OE, Hamad A, Fox J, DeFrank A, Gilley R, Dawson F, Sykes B, Underwood TJ, Naylor RC, Briggs AA, Lassiter CS, Bell WE, Turner JE. Estrogen prevents cardiac and vascular failure in the 'listless' zebrafish (*Danio rerio*) developmental model. General and Comparative Endocrinology. 2013;**189**:33-42

[26] Murcia V, Johnson L, Baldasare M, Pouliot B, McKelvey J, Barbery B, Lozier J, Bell WE, Turner JE. Estrogen, nitric oxide and dopamine interactions in the zebrafish "listless" model of locomotor dysfunction. Toxics. 2016;**4**:24. DOI: 10.3390/toxis4040024

[27] White RM, Sessa A, Burke C, Bowman T, LeBlanc J, Ceol C, Bourque C, Zon LI. Transparent adult zebrafish as a tool for in vivo transplantation analysis. Cell Stem Cell. 2008; **2**:183-189

[28] Houser A, McNair C, Piccinini R, Luxhoj A, Bell WE, Turner JE. Effects of estrogen on the neuromuscular system in the embryonic zebrafish (*Danio rerio*). Brain Research. 2011; **1381**:106-116

[29] Giacomini NJ, Rose B, Kobayashi K, Guo S. Antipsychotics produce locomotor impairment in larval zebrafish. Neurotoxicology and Teratology. 2006;**28**:245-250

[30] Sheng D, Qu D, Kwok KH, Ng SS, Lim AY, Aw SS, Lee CW, Sung WK, Tan K, Lufkin T, Jesuthasan S, Sinnakaruppan M, Liu J. Deletion of the WD40 domain of LRRK2 in Zebrafish causes parkinsonism-like loss of neurons and locomotive defect. PLoS Genetics. 2010;**6**:e1000914

[31] Irons TD, Kelly PE, Hunter DL, Macphail RC, Padilla S. Acute administration of dopaminergic drugs has differential effects on locomotion in larval fish. Pharmacology, Biochemistry, and Behavior. 2013;**103**:702-813

[32] Stednitz SJ, Freshner B, Shelton S, Shen T, Black D, Gahtan E. Selective toxicity of L-DOPA to dopamine transporter-expressing neurons and locomotor behavior in zebrafish larvae. Neurotoxicology and Teratology. 2015;**52**:51-56

[33] Del-Bel E, Padovan-Neto FE, Bortolanza M, Tumaas V, Junior AA, Raisman-Vozari R, Prediger RD. Nitric oxide, a new player in l-dopa-induced dyskinesia. Frontiers in Bioscience. 2015;1:168-192

[34] Saili KS, Corvi MM, Weber DN, Patel AU, Da SR, Przybyla J, Anderson KA, Tanguay RL. Neurodevelopmental low-dose bisphenol a exposure leads to early life-stage hyperactivity and learning deficits in adult zebrafish. Toxicology. 2012;291:83-92

[35] Komada M, Itoh S, Kawachi K, Kanawa N, Ikeda Y, Nagao T. Newborn mice exposed prenatally to bisphenol A show hyperactivity and defective neocortical development. Toxicology. 2014;323:51-60

[36] Nojima K, Takata T, Masuno H. Prolonged exposure to a low-dose of bisphenol A increases spontaneous motor activity in adult male rats. The Journal of Physiological Sciences. 2013; 63:311-315

9

Zebrafish Model of Cognitive Dysfunction

Brandon Kar Meng Choo and Mohd. Farooq Shaikh

Abstract

Cognitive dysfunction is an impairment in one or more of the six cognitive domains (complex attention, executive function, learning and memory, language, perceptual motor and social cognition). The effect of pharmacological interventions can be studied using animal models of cognitive dysfunction, which are typically split into pharmacological, developmental and genetic models. Rodents are the most commonly used animal species for modelling cognitive dysfunction, although multiple models and test locations are often recommended to improve validity. Researchers thus unfortunately need to balance the validity of their experimental designs with financial, logistical and cost constraints. Zebrafish could be the answer to this conundrum as one of their many advantages over rodents is their high breeding rate which makes high-throughput screening more feasible and thus increases cost-effectiveness. The popularity of zebrafish has been increasing in recent times, as measured by the increasing number of zebrafish research publications. It is thus unsurprising that several zebrafish models of cognitive dysfunction have already been developed, together with zebrafish tests designed to measure zebrafish cognitive performance. Future research will undoubtedly lead to the development of new zebrafish models of cognitive dysfunction, as well as validate current ones to pave the way for widespread adoption.

Keywords: zebrafish, cognition, animal model, cognitive dysfunction, drug discovery

1. Introduction to cognitive dysfunction

Cognitive dysfunction is an impairment in one of the six cognitive domains as defined by the Diagnostic and Statistical Manual of Mental Disorders, Fifth Edition (DSM-5). The six cognitive domains are complex attention, executive function, learning and memory, language, perceptual motor and social cognition [1]. Cognitive disorders are a category of mental health disorders and are officially termed 'neurocognitive disorders' by the DSM-5. There is a key difference between cognitive impairment in neurocognitive disorders and in neurodevelopmental disorders; in that, the former is acquired, whereas the latter develops at birth or shortly thereafter. Neurocognitive disorders are broadly divided into mild neurocognitive disorders and major neurocognitive disorders, which are mostly made up of the dementias [1]. The difference between the two categories is innately arbitrary, as the diagnostic criteria rely on determining the degree of cognitive and functional impairment in the patient. The DSM-5 itself explicitly acknowledges that precise thresholds are difficult to determine because the cognitive and functional impairments associated with neurocognitive disorders exist in a spectrum. An impairment in one cognitive domain is sufficient for the diagnosis of a neurocognitive disorder, except in the case of Alzheimer's disease, whereby impairment of memory and one other domain is required [2]. Although Alzheimer's disease is typically responsible for the majority of neurocognitive disorders, among the medical conditions which can also affect mental functions are frontotemporal degeneration, Huntington's disease, Lewy body disease, traumatic brain injury, Parkinson's disease, prion disease and dementia/neurocognitive issues due to HIV infection [2, 3].

Treatments for cognitive dysfunction are broadly divided into psychopharmacological interventions and behaviour-based cognitive remediation, though both methods can be combined to supplement one another. Cognitive remediation focuses on teaching patients cognitive skills such as thinking and problem solving, which are typically taught in formal education or through real-world activities which the patient may not have had the opportunity to experience due to their condition [4]. In contrast, psychopharmacological interventions such as antipsychotics primarily aim to treat or manage the underlying cause of cognitive dysfunction but also improve cognition as they are usually capable of affecting a variety of central neuroreceptors [5]. Examples of antipsychotics which improve cognition include serotonin receptor antagonists such as clozapine [6] and α_{2C}-adrenoceptor antagonists such as iloperidone [7].

2. The animal model of cognitive dysfunction

While studies on behavioural-based cognitive remediation are perhaps best done in humans, the effect of pharmacological interventions can be studied using animal models of cognitive dysfunction. The following subsection aims to provide a brief overview of the six cognitive domains defined by the DSM-5 and the animal models which can be used to model a dysfunction in each domain. Animal models of cognitive dysfunction are typically split into three major categories, namely, the pharmacological, developmental and genetic models. Pharmacological models utilize treatments such as dopamine and serotonin agonists, to produce effects which mimic cognitive dysfunction. Developmental models induce cognitive dysfunction in young animals via brain lesions or by disrupting the normal maturation process by exposing the animals to certain compounds such as L-nitroarginine. Finally, genetic models produce animals with cognitive dysfunctions through either inbreeding or genetic modification [8].

2.1. Complex attention

The cognitive domain of complex attention consists of sustained attention, divided attention, selective attention and information processing speed [9]. As the aptly named attention deficit hyperactivity disorder (ADHD) is likely to be the first disorder affecting this domain which comes to mind, it is hardly surprising that animal models of a dysfunction in complex attention are usually also animal models of ADHD. One such animal model of complex attention is spontaneously hypertensive rats which exhibit all the characteristic behaviours of ADHD, including impaired sustained attention [10]. Animal tests have also been developed to study sustained attention, such as the '5-choice serial reaction time task' as well as an operant task which requires rats to detect and discriminate between signals and non-signals [11]. The '5-choice serial reaction time task' can also be used as a measure of information processing speed [12].

2.2. Executive function

The cognitive domain of executive function consists of planning, decision-making, working memory, responding to feedback/error correction, overriding habits and mental flexibility [9]. Executive function is typically associated with the prefrontal cortex and is believed to be significantly more complex in primates as compared to rodents [12]. However, the use of rodent models of executive function should not be entirely discounted for the same reasons that they have become widely used today as animal models. These reasons include being a more cost effective and simpler system for study, but one that still possesses many of the complex characteristics which are of interest [13]. There is also evidence to believe that rats are also capable of executive functions such as decision-making and that homologous regions within the prefrontal cortex of primates and rats also have parallel cognitive functions [13]. Among the aspects of executive function which have been replicated using animal models is working memory, using tasks such as the spatial delayed response and spatial search tests. Decision-making has also been modelled using animals by utilising behavioural tasks (reversal learning, reinforcer devaluation and delay discounting), which require flexible adaptation in response to a changing environment [14]. As a decline in executive function is a characteristic of the ageing process, aged rats can also be used to simulate age-related cognitive decline [12].

2.3. Learning and memory

The cognitive domain of learning and memory consists of immediate memory and recent memory (including free recall, cued recall and recognition memory) [9]. Inbred 'senescence accelerated' mice display spontaneously occurring age-related learning and memory defects, as well as difficulty in acquiring new behaviours [15]. Animal models of Alzheimer's disease such as the PDAPP transgenic mice can also be used as models of learning and memory dysfunction as they display memory impairments in object recognition tasks and also have difficulty in learning tasks [16]. Other methods of producing animal models with impaired learning and memory include the genetic modification of genes such as the tuberous sclerosis genes [17] and causing brain injury to the animals in the form of mild trauma or concussions [18]. Several tests designed to assess animal learning and memory include the Morris water maze [19] and passive avoidance tasks which utilise apparatus such as the elevated T-maze [20].

2.4. Language

The cognitive domain of language consists of expressive language (including naming, fluency, grammar and syntax) and receptive language [9]. Currently, no animal model has been produced which mimics an acquired language dysfunction in humans. However, a rodent model of developmental language disability has already been developed from an observation that humans with impaired language development also have difficulty in processing rapidly presented auditory information [21]. It is the authors' hope that animal models of language dysfunction will be developed in the future once it becomes possible to teach animals the languages used by humans or for humans to learn animal languages.

2.5. Perceptual motor

The perceptual motor cognitive domain consists of construction and visual perception [9]. No animal models specifically meant to replicate a perceptual motor dysfunction have been found in literature. This could possibly be because perceptual motor skills rely on integrating information from a variety of sensory inputs and are thus exceedingly challenging to replicate in animal studies. As the name suggests, perception is a major component of this domain but may be difficult to assess in animals. Perceptual motor skills may also involve fine motor tasks, which may be difficult to teach all but non-human primates [22].

2.6. Social cognition

The cognitive domain of social cognition consists of recognition of emotions, theory of mind and behavioural regulation [9]. Animal models of social cognition have been produced, such as the oxytocin receptor null mice model of autism which displays impaired social memory [23] said to be analogous to human social cognition deficits [24]. Valproic acid-exposed rodents have also been used to model the behavioural characteristics of autism spectrum disorder, including a dysfunction of social cognition, as they tend to avoid interacting socially [25]. Social cognition can also be assessed through rodent behaviour tests, the degree of social interaction and so forth, although there is debate on whether these measures actually assess social behaviour or social cognition. However, certain social constructs such as 'theory of mind' will require the use of non-human primates or possibly another animal species besides rodents [24].

2.7. An overview of the current animal models for modelling cognitive dysfunction

Even from the brief overview above, it is clear that animals have been an invaluable tool for modelling dysfunctions in four of the six cognitive domains, though modelling a dysfunction in the language and perceptual motor cognitive domains is currently proving to be a challenge. Rodents are by far the most commonly used animal species for modelling cognitive dysfunction, though other species such as non-human primates, felines and canines have been used as well [26]. Whilst there is great discord on the validity of rodent models in comparison to the more complex non-human primate models, rodents are preferred by the vast majority of laboratories worldwide due to economical, logistical and ethical constraints regarding the use of non-human primates [26]. However, rodent models have been criticised for lacking in predictability when translating results into human clinical trials [27, 28]. One possible way of improving the predictability of rodent models is to use more than one model and to replicate the study at different locations [27]. While such an experimental design may improve the validity of rodent models of cognitive dysfunction, the financial costs will undoubtedly rise in tandem with the number of models and test locations used, in addition to increasing logistical and spatial requirements. Thus, researchers will unfortunately need to balance the validity of their experimental designs with financial, logistical and cost constraints.

3. Zebrafish as an emerging animal model of cognitive dysfunction

The solution to the problem of validity versus practicality could be as simple as replacing rodents with another animal species such as zebrafish (*Danio rerio*), which are becoming increasingly popular [29] (**Figure 1**) due to their high breeding rate, which in turn makes high-throughput screening feasible. This is because the oviparous zebrafish breed continuously throughout the year and have relatively short generation times of between 3 and 5 months [30]. Zebrafish offspring also quickly mature ex utero, gaining vision and the ability to swim freely as well as feed, all within 72 hours. Larval zebrafish in their early life can survive in just 50 µl of solution, enabling the use of microtitre plates for high-throughput screening to discover compounds with a certain desired effect. In addition to being relatively small, zebrafish larvae are also translucent, and thus in vivo imaging techniques such as fluorescent reporters may be used to monitor the progression of diseases at the cellular and subcellular levels [31]. Dissolving the compounds to be tested directly in the tank water is also an option with zebrafish, eliminating the necessity of performing invasive procedures such as injections. Zebrafish are thought to absorb substances dissolved in the tank water through the gastrointestinal tract and also via the transdermal route in the case of immature zebrafish. These routes of absorption are possible

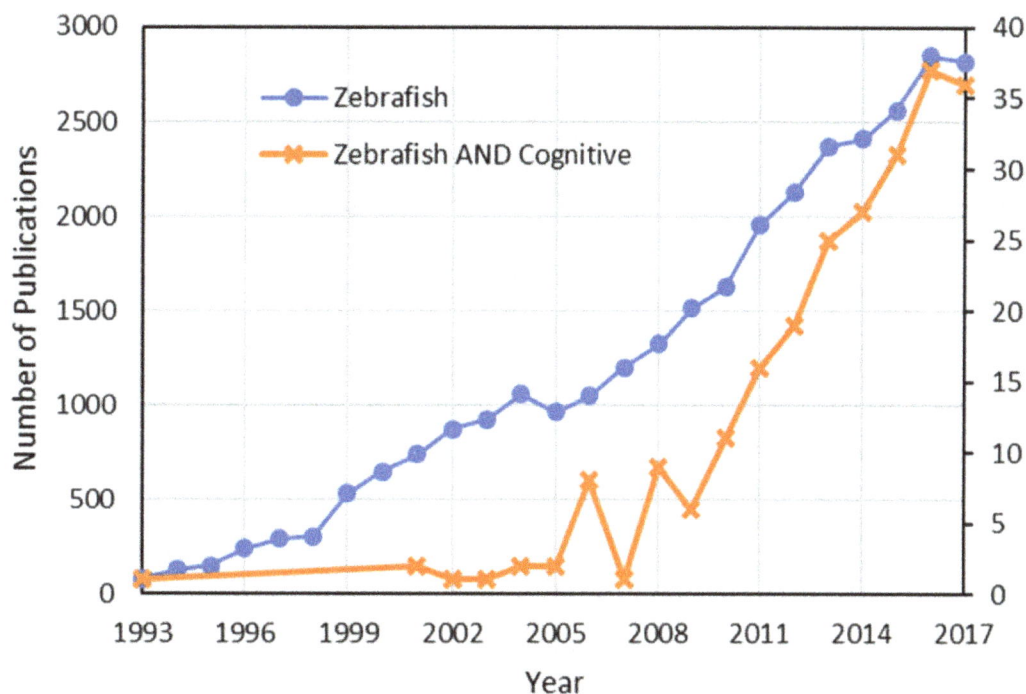

Figure 1. Publications with the keyword 'Zebrafish' versus publications with the keywords 'Zebrafish AND Cognitive', as indexed by PubMed from the year 1993 (first appearance of articles with the keywords 'Zebrafish AND Cognitive') to 2017.

in immature zebrafish as they start swallowing from their third day of life and do not develop scales until they are several weeks old. Despite zebrafish being fish and hence more removed from humans in an evolutionary sense as compared to mammalian rodents, their genes are nonetheless around 75% homologous to human genes. This makes identifying the human orthologues present in zebrafish relatively simple as both human and zebrafish genomes have already been fully sequenced. Additionally, constructs of RNA, protein or DNA can be injected into embryos at an early stage of development to modify their gene and hence protein expression. Morpholinos are modified oligonucleotides which may be inserted into zebrafish embryos via an injection to decrease the expression of selected genes. This method enables as many as 500 knockout zebrafish embryos to be swiftly and effortlessly produced over a period of several hours [32, 33]. The blood-brain barrier in zebrafish is also tight junction based and highly permeable to macromolecules, meaning that zebrafish will be extremely responsive to test compounds [34]. Anatomy wise, zebrafish brains also have similarities to human brains in that both possess defined forebrain, midbrain and hindbrain. Both humans and zebrafish also have a diencephalon, telencephalon and cerebellum, as well as peripheral nervous system with motor, sensory and autonomic components [35]. Zebrafish also exhibit 'higher' behaviours and show integrated neural functions such as memory, conditioned responses and social behaviour [35]. All these aspects make zebrafish an attractive animal model for research into cognitive dysfunction due to their cost-effectiveness, scalability and similarity to humans as compared to several other animal models (**Table 1**). In addition, pharmacologic, developmental and genetic animal models of cognitive dysfunction can be easily produced using zebrafish and on a much larger scale as compared to rodents due to their higher breeding rate. The 'higher' behaviours and cognitive ability exhibited by zebrafish will also enable the use of cognitive and behaviour tests to assess cognitive dysfunction, much like those used for rodent models.

3.1. The progress towards developing a zebrafish model of cognitive dysfunction

While the concept of using zebrafish as an animal model may be relatively new as compared to other more established animal models such as rodents, zebrafish models of cognitive dysfunction and zebrafish cognitive tests have already been developed. While rodent pharmacologic and developmental models of cognitive dysfunction have parallels in zebrafish, no genetic models have been developed thus far. Zebrafish have also been found to display an age-related decline in cognitive dysfunction [36], similar to that found in other animal models such as rodents. Unfortunately, most zebrafish cognitive dysfunction models excluding those which involve unconditioned or reflexive behaviours may need to be carried out in adults as the neural system of zebrafish is still immature at the larval stage [37].

3.1.1. Complex attention

In the case of the cognitive domain of complex attention, there is no attention task for zebrafish which is widely accepted, and thus validated zebrafish behaviour tasks for measuring learning are used to infer attention [38]. However, the march of progress is relentless, and tasks have already been developed to test complex attention, such as the 'virtual object recognition test' [39] and the '3-choice serial reaction time task' [40] for adult zebrafish. It is

Animal model	Canines/felines	Non-human primates	Rodents	Zebrafish
Upkeep cost per animal	$$$	$$$$	$$	$
Space required	●●●○	●●●●	●●○○	●○○○
Breeding rate	●●○○	●○○○	●●●○	●●●●
Feasibility of high-throughput screening	●○○	●○○	●●○	●●●
Anatomical similarity to humans	●●○	●●●	●●○	●○○
Genetic similarity to humans	●●○	●●●	●○○	●○○

The '$' symbols represent the relative upkeep cost for the four animal models, with a greater number of '$' symbols signifying a greater cost. The number of shaded circles as compared to unshaded circles serves as a relative comparison for the different attributes between the four animal models, with the superlative having all circles shaded.

Table 1. A comparison of the current animal models of cognitive dysfunction.

interesting to note that both zebrafish tasks are adapted from the tasks used in rodents, which bode well for the prospects of both tests being widely accepted in the future for the testing of complex attention in zebrafish.

3.1.2. Executive function

For the cognitive domain of executive function, reversal learning tasks have been used to demonstrate the behavioural flexibility of adult zebrafish, which is associated with executive function [37]. However, it should be noted that executive function also includes aspects of working memory and feedback/error correction [9]. Thus, zebrafish models meant to assess the cognitive domains of complex attention as well as learning and memory could also be used to assess executive function, depending on the working definition regarding the scope which executive function covers.

3.1.3. Learning and memory

For the learning and memory cognitive domain, the behaviour tasks which are used to assess this cognitive domain include the condition place preference, predator avoidance, T-maze, plus maze, three-compartment zebrafish maze and the three-choice discrimination tests [38] for adult zebrafish. The pharmacological method of inducing cognitive dysfunction has also been used to produce adult zebrafish which have deficits in learning and memory, by treating them with compounds such as antiepileptic drugs [41] or scopolamine [42]. Larval zebrafish may also be used to model nonassociative learning as they display short- and long-term habituation to visual and acoustic stimuli, which leads to the suppression of characteristic manoeuvre responses to these stimuli [43].

3.1.4. Language

In the case of the language cognitive domain, zebrafish models of dysfunction in this domain are also currently absent from literature, possibly due to the reasons previously discussed.

3.1.5. Perceptual motor

In the case of the perceptual motor cognitive domain, zebrafish models of dysfunction in this domain are also currently absent from literature, possibly due to the reasons previously discussed.

3.1.6. Social cognition

Lastly, the cognitive domain of social cognition is measured in adult zebrafish by exploiting its natural shoaling behaviour. The shoaling may be achieved using live zebrafish or by presenting computer-generated images of zebrafish and quantifying the resulting shoaling behaviour using parameters such as the distance between the test zebrafish and the other shoaling members [44]. Pharmacological methods of inducing social cognition dysfunction include treatment with dizocilpine [45]. Ethanol can also be used as a developmental model of social cognition dysfunction, as zebrafish embryos exposed to ethanol develop social cognition dysfunction in adulthood [46].

3.2. An overview of several cognitive dysfunction tests developed for zebrafish

3.2.1. Three-choice discrimination

3.2.1.1. Purpose

- Evaluates all five of Bushnell's five categories of attention (orienting, expectancy, stimulus differentiation, sustained attention and parallel processing) in an animal, when modified to include additional stimuli as potential distractions [38, 47].

- Requires zebrafish to be able to orient themselves as well as being able to anticipate the result of an action.

3.2.1.2. Procedure

- A zebrafish is trained to swim from a start chamber into one of three chambers using food as a reinforcement.

- The correct chamber which contains food is illuminated with white light, whereas the empty incorrect chambers are left dark.

3.2.1.3. Critical assessment

- The zebrafish must overcome its preference for dark places in order to select the correct chamber, and this demonstrates that the zebrafish is learning rather than acting solely on instinct.

- The addition of a third chamber is an improvement over the T-maze as it reduces the chance level from 50 to 33%.

3.2.2. Three-choice serial reaction time

3.2.2.1. Purpose

- Evaluates the ability of a zebrafish to learn a complex behavioural task [40, 48].

- Demonstrates the capacity of zebrafish for learning and memory.

3.2.2.2. Procedure

- A zebrafish is first trained to approach the response aperture of a tank when it is lit by a stimulus light, using food as a reinforcement.

- The trained zebrafish is then placed in a tank with three response apertures and given a brief period to nose poke the lit aperture to receive reinforcement in the form of food.

3.2.2.3. Critical assessment

- The correlation with the rodent equivalent from which this test is derived from is unclear due to anatomical differences between the two species.

- It is also unclear if the stimulus light is visible from all areas of the test tank and thus a failure of the zebrafish to see the stimulus light could be incorrectly perceived as a deficit in learning and memory.

- The amount of time that the zebrafish is given to eat the food reinforcement should also be carefully chosen to allow sufficient time for the zebrafish to finish the food.

3.2.3. Three-compartment zebrafish maze

3.2.3.1. Purpose

- Evaluates spatial discrimination learning in zebrafish and also demonstrates avoidance discrimination learning in zebrafish [38, 49].

3.2.3.2. Procedure

- A zebrafish is first placed in the middle of a three-chambered tank, with partitions at either end of the central chamber.

- After a minute, the partitions are lifted, and the zebrafish is allowed to swim into either the left or right chambers.

- If the zebrafish swims into the chamber designated as the 'wrong' side, the partition is pushed until it is within 1 cm from the end of the tank, in order to confine the zebrafish and 'punish' it.

- After 10 seconds, the zebrafish is allowed to return to the central chamber, and the protocol can be repeated as needed.

3.2.3.3. Critical assessment

- Zebrafish can reliably learn and remember the response contingencies when this behavioural task is repeated multiple times.

3.2.4. Plus maze

3.2.4.1. Purpose

- Evaluates a zebrafish's ability to learn and remember the association between a single visual cue and a food reward (simple associative learning), as well as the location of the food reward (spatial learning) [38, 50].

3.2.4.2. Procedure

- The plus maze is placed on a rotating circular platform, and a zebrafish is first transferred into the middle of the plus maze (four-armed radial maze), which is bounded by start box to prevent the zebrafish from entering the arms before the start of the experiment.

- The zebrafish then undergoes habituation (food provided in all arms) and shaping trials (food provided only in certain arms next to a visual cue, such as a red plastic card).

- The final training step involves the use of paired and unpaired groups whereby the food is provided together with or independently of the visual cue.

- Zebrafish learning can then be tested by providing only the visual cue with the food being inaccessible (associative learning) or by providing the food in a fixed location relative to external visual cues such as room equipment, without any visual cues in the maze itself (spatial learning).

- Parameters such as the time spent in the target arm and the number of entries into the target arm versus other arms are quantified to assess zebrafish learning and memory.

3.2.4.3. Critical assessment

- The four distinct spatial locations provided by the plus maze is again an improvement over the T-maze.

- The plus maze provides stimuli in a predictable way and is thus unsuitable for measuring sustained attention.

- It is also possible that the spatial learning task is accomplished by the zebrafish via non-spatial strategies such as by remembering a single salient cue next to the target location.

3.2.5. Reversal of learning

3.2.5.1. Purpose

- Evaluates the ability of a zebrafish to adapt or shift their response strategy when presented with changing environmental contingencies [37].

3.2.5.2. Procedure

- Commonly involves first training a zebrafish to discriminate between two stimuli such as different coloured lights and only reinforcing responses to one stimulus.

- Once the zebrafish reaches a certain response criterion, the contingency is reversed so that the other stimulus is reinforced and the previously 'correct' stimulus is not.

- Subsequent extensions to this test include changing the colour of the stimuli (intra-dimensional shift) or adding a third dimension such as shape, before the reversal (extradimensional shift).

3.2.5.3. Critical assessment

- Zebrafish follow a similar pattern of improvement as mammals when subjected to multiple reversals and interdimensional shifts as they require increasingly fewer trials to reach the response criterion.

3.2.6. Shoaling behaviour

3.2.6.1. Purpose

- Evaluates shoaling behaviour in zebrafish, which is one aspect of zebrafish social behaviour [44].

3.2.6.2. Procedure

- A group of around ten zebrafish are first habituated together in the experimental tank on several occasions before use.

- Each of the habituated zebrafish can then be placed individually in an experimental tank without any stimuli, in order to measure their baseline behaviour.

- Shoaling behaviour can subsequently be induced using methods such as:

  ○ Placing live stimulus fish inside/outside the experimental tank.

  ○ Displaying animated zebrafish images.

- Shoaling behaviour is recorded and subsequently quantified by video tracking software to determine behaviour parameters such as distance to the stimulus, distance to the bottom of the tank and angular velocity.

3.2.6.3. Critical assessment

- Zebrafish are diurnal like humans and thus have good vision, which helps the face validity of this test.

- Electronic devices meant for humans can also be used for this zebrafish test, which increases cost-effectiveness.

- It is not known if virtual fish are truly equivalent to live fish in inducing zebrafish social behaviour as virtual fish do not interact with the experiment fish.

- Other social behaviours such as aggression and reproduction have not been studied using this model.

3.2.7. T-maze

3.2.7.1. Purpose

- Evaluates a zebrafish's ability associate visual stimuli with a food reward (similar to the plus maze) [51].

3.2.7.2. Procedure

- Similar to the plus maze, but the T-maze only possesses two arms rather than four arms.

- Different coloured or patterned sleeves could be fitted to the arms or goal boxes instead of lights.

3.2.7.3. Critical assessment

- Similar to the plus maze, although the T-maze only contains two chambers and thus fewer distinct spatial locations.

3.2.8. Virtual object recognition

3.2.8.1. Purpose

- Evaluates an animal's attention when presented with novel stimuli [39].

- Zebrafish can identify shape information independently of motion information.

3.2.8.2. Procedure

- A zebrafish is placed in a transparent tank, and two displays are used to display either static or dynamic identical images on both ends of the tank.

- After a 10-minute familiarisation trial, the zebrafish is returned to the home tank and later subjected to a novel shape recognition trial (one familiar image, one novel image).

- Recognition time (seconds) is measured when the zebrafish approaches the display area and orients its head towards the display.

3.2.8.3. Critical assessment

- Testing sequence is fairly rapid, and no animal training is required, other than the initial exposure session.

3.2.9. Zebrafish larvae visual/acoustic stimuli habituation

3.2.9.1. Purpose

- Evaluates zebrafish larvae nonassociative learning as the test requires a degree of memory storage and retrieval [43].

- Zebrafish larvae show long-term habituation to visual stimuli and short-term habituation to acoustic stimuli, as demonstrated by a decline in the rate of characteristic movements in response to repeated exposure to visual or acoustic stimuli.

3.2.9.2. Procedure

- Visual stimulation is provided by first equilibrating a zebrafish larva to a uniformly lit testing chamber and then abruptly extinguishing the light, which triggers a unique turning behaviour termed the O-bend:

  - The training procedure is then repeated at desired intervals in either a massed or spaced fashion, before returning the zebrafish to a holding tank.

  - The visual stimulation can then be repeated after a desired amount of time to determine how much of the habituation training remains.

- Acoustic stimulation is provided by exposing a zebrafish larva to a series of acoustic stimuli with varying intensities and intervals between each stimulus, to trigger a characteristic kinematic startle response termed 'short-latency C-start':

  - After a short resting period of several minutes, the protocol is repeated to determine the degree of habituation.

3.2.9.3. Critical assessment

- Being able to use larval zebrafish allows for high-throughput screening methods.

- The characteristic responses are relatively simple and may not be suitable for testing higher cognitive functions.

- Memory retention time also appears to be relatively short, ranging from minutes in short-term habituation to hours in long-term habituation. Thus, zebrafish larvae may not be suitable for testing learning and memory over longer periods of time.

4. Conclusions and future directions

In conclusion, zebrafish have the potential to replace rodents as the most widely used model of cognitive dysfunction, mainly due to their higher cost-effectiveness, and may have already begun doing so as several zebrafish models of cognitive dysfunction have already been developed. Future areas of research into the zebrafish model of cognitive dysfunction could focus on producing new models of cognitive dysfunction to complement the existing models, as well as to expand the scope of research into cognitive dysfunction that is possible using zebrafish. Already existing models should also be further examined to validate them in hopes that in the near future, certain tasks will become widely accepted tests of cognitive dysfunction in zebrafish. Another potential area of research is the development and characterisation of mutant zebrafish strains to produce genetic models of cognitive dysfunction in zebrafish.

Conflict of interest

The authors declare that the research was conducted in the absence of any commercial or financial relationships that could be construed as a potential conflict of interest.

Author details

Brandon Kar Meng Choo and Mohd. Farooq Shaikh*

*Address all correspondence to: farooq.shaikh@monash.edu

Neuropharmacology Research Laboratory, Jeffrey Cheah School of Medicine and Health Sciences, Monash University Malaysia, Bandar Sunway, Malaysia

References

[1] American Psychiatric Association. Diagnostic and Statistical Manual of Mental Disorders (DSM-5®): American Psychiatric Pub; 2013

[2] Simpson JR. DSM-5 and neurocognitive disorders. Journal of the American Academy of Psychiatry and the Law Online. 2014;**42**(2):159-164

[3] Sachdev PS, Blacker D, Blazer DG, Ganguli M, Jeste DV, Paulsen JS, et al. Classifying neurocognitive disorders: The DSM-5 approach. Nature Reviews Neurology. 2014;**10**(11): 634-642

[4] Medalia A, Lim R. Treatment of cognitive dysfunction in psychiatric disorders. Journal of Psychiatric Practice®. 2004;**10**(1):17-25

[5] Keefe RS, Silva SG, Perkins DO, Lieberman JA. The effects of atypical antipsychotic drugs on neurocognitive impairment in schizophrenia: A review and meta-analysis. Schizophrenia Bulletin. 1999;**25**(2):201-222

[6] Meltzer HY, Sumiyoshi T. Does stimulation of 5-HT 1A receptors improve cognition in schizophrenia? Behavioural Brain Research. 2008;**195**(1):98-102

[7] Kalkman HO, Loetscher E. α 2C-Adrenoceptor blockade by clozapine and other antipsychotic drugs. European Journal of Pharmacology. 2003;**462**(1):33-40

[8] Geyer MA, Moghaddam B. Animal models relevant to schizophrenia disorders. Neuropsychopharmacology: The Fifth Generation of Progress. 2002:689-701

[9] Goodkin K, Fernandez F, Forstein M, Miller EN, Becker JT, Douaihy A, et al. A perspective on the proposal for neurocognitive disorder criteria in DSM-5 as applied to HIV-associated neurocognitive disorders. Neuropsychiatry. 2011;**1**(5):431-440

[10] Russell VA, Sagvolden T, Johansen EB. Animal models of attention-deficit hyperactivity disorder. Behavioral and Brain Functions. 2005;**1**(1):9

[11] Sarter M, Givens B, Bruno JP. The cognitive neuroscience of sustained attention: Where top-down meets bottom-up. Brain Research Reviews. 2001;**35**(2):146-160

[12] Bizon JL, Foster TC, Alexander GE, Glisky EL. Characterizing cognitive aging of working memory and executive function in animal models. Frontiers in Aging Neuroscience. 2012;**4**:19

[13] Kesner RP, Churchwell JC. An analysis of rat prefrontal cortex in mediating executive function. Neurobiology of Learning and Memory. 2011;**96**(3):417-431

[14] Chudasama Y. Animal models of prefrontal-executive function. Behavioral Neuroscience. 2011;**125**(3):327

[15] Yagi H, Katoh S, Akiguchi I, Takeda T. Age-related deterioration of ability of acquisition in memory and learning in senescence accelerated mouse: SAM-P/8 as an animal model of disturbances in recent memory. Brain Research. 1988;**474**(1):86-93

[16] Dodart J-C, Bales KR, Gannon KS, Greene SJ, DeMattos RB, Mathis C, et al. Immunization reverses memory deficits without reducing brain Aβ burden in Alzheimer's disease model. Nature Neuroscience. 2002;**5**(5):452-457

[17] Ehninger D, Han S, Shilyansky C, Zhou Y, Li W, Kwiatkowski DJ, et al. Reversal of learning deficits in a Tsc2(+/−) mouse model of tuberous sclerosis. Nature Medicine. 2008;**14**(8):843-848

[18] Mouzon B, Chaytow H, Crynen G, Bachmeier C, Stewart J, Mullan M, et al. Repetitive mild traumatic brain injury in a mouse model produces learning and memory deficits accompanied by histological changes. Journal of Neurotrauma. 2012;**29**(18):2761-2773

[19] D'Hooge R, De Deyn PP. Applications of the Morris water maze in the study of learning and memory. Brain Research Reviews. 2001;**36**(1):60-90

[20] Viana MB, Tomaz C, Graeff FG. The elevated T-maze: A new animal model of anxiety and memory. Pharmacology Biochemistry and Behavior. 1994;**49**(3):549-554

[21] Clark MG, Rosen GD, Tallal P, Fitch RH. Impaired processing of complex auditory stimuli in rats with induced cerebrocortical microgyria: An animal model of developmental language disabilities. Journal of Cognitive Neuroscience. 2000;**12**(5):828-839

[22] Finney GR. Perceptual-motor dysfunction. Continuum: Lifelong Learning in Neurology. 2015;**21**(3, Behavioral Neurology and Neuropsychiatry):678-689

[23] Sala M, Braida D, Lentini D, Busnelli M, Bulgheroni E, Capurro V, et al. Pharmacologic rescue of impaired cognitive flexibility, social deficits, increased aggression, and sei-

zure susceptibility in oxytocin receptor null mice: A neurobehavioral model of autism. Biological Psychiatry. 2011;**69**(9):875-882

[24] Green MF, Olivier B, Crawley JN, Penn DL, Silverstein S. Social cognition in schizophrenia: Recommendations from the measurement and treatment research to improve cognition in schizophrenia new approaches conference. Schizophrenia Bulletin. 2005; **31**(4):882-887

[25] Foley AG, Gannon S, Rombach-Mullan N, Prendergast A, Barry C, Cassidy AW, et al. Class I histone deacetylase inhibition ameliorates social cognition and cell adhesion molecule plasticity deficits in a rodent model of autism spectrum disorder. Neuropharmacology. 2012;**63**(4):750-760

[26] Cenci MA, Whishaw IQ, Schallert T. Animal models of neurological deficits: How relevant is the rat? Nature Reviews Neuroscience. 2002;**3**(7):574-579

[27] Jucker M. The benefits and limitations of animal models for translational research in neurodegenerative diseases. Nature Medicine. 2010;**16**(11):1210-1214

[28] Zahs KR, Ashe KH. 'Too much good news'–are Alzheimer mouse models trying to tell us how to prevent, not cure, Alzheimer's disease? Trends in Neurosciences. 2010;**33**(8):381-389

[29] Geisler R, Köhler A, Dickmeis T, Strähle U. Archiving of zebrafish lines can reduce animal experiments in biomedical research. EMBO Reports. 2016;**18**:1-2

[30] Spitsbergen JM, Kent ML. The state of the art of the Zebrafish model for toxicology and Toxicologic pathology research—Advantages and current limitations. Toxicologic Pathology. 2003;**31**(1_suppl):62-87

[31] Paquet D, Bhat R, Sydow A, Mandelkow E-M, Berg S, Hellberg S, et al. A zebrafish model of Tauopathy allows in vivo imaging of neuronal cell death and drug evaluation. The Journal of Clinical Investigation. 2009;**119**(5):1382-1395

[32] Baraban S, Taylor M, Castro P, Baier H. Pentylenetetrazole induced changes in zebrafish behavior, neural activity and c-fos expression. Neuroscience. 2005;**131**(3):759-768

[33] Berghmans S, Hunt J, Roach A, Goldsmith P. Zebrafish offer the potential for a primary screen to identify a wide variety of potential anticonvulsants. Epilepsy Research. 2007;**75**(1):18-28

[34] Stewart AM, Desmond D, Kyzar E, Gaikwad S, Roth A, Riehl R, et al. Perspectives of zebrafish models of epilepsy: What, how and where next? Brain Research Bulletin. 2012;**87**(2):135-143

[35] Dooley K, Zon LI. Zebrafish: A model system for the study of human disease. Current Opinion in Genetics & Development. 2000;**10**(3):252-256

[36] Yu L, Tucci V, Kishi S, Zhdanova IV. Cognitive aging in zebrafish. PLoS One. 2006;**1**(1):e14

[37] Parker MO, Brock AJ, Walton RT, Brennan CH. The role of zebrafish (*Danio rerio*) in dissecting the genetics and neural circuits of executive function. Frontiers in Neural Circuits. 2013;**7**:63

[38] Echevarria DJ, Jouandot DJ, Toms CN. Assessing attention in the zebrafish: Are we there yet? Progress in Neuro-Psychopharmacology and Biological Psychiatry. 2011;**35**(6):1416-1420

[39] Braida D, Ponzoni L, Martucci R, Sala M. A new model to study visual attention in zebrafish. Progress in Neuro-Psychopharmacology and Biological Psychiatry. 2014;**55**:80-86

[40] Parker M, Millington ME, Combe F, Brennan C. Development and implementation of a three-choice serial reaction time task for zebrafish (*Danio rerio*). Behavioural Brain Research. 2011:73-80

[41] Kundap UP, Kumari Y, Othman I, Shaikh MF. Zebrafish as a model for epilepsy-induced cognitive dysfunction: A pharmacological, biochemical and behavioral approach. Frontiers in Pharmacology. 2017;**8**(515)

[42] Richetti SK, Blank M, Capiotti KM, Piato AL, Bogo MR, Vianna MR, et al. Quercetin and Rutin prevent scopolamine-induced memory impairment in zebrafish. Behavioural Brain Research. 2011;**217**(1):10-15

[43] Wolman MA, Jain RA, Liss L, Granato M. Chemical modulation of memory formation in larval zebrafish. Proceedings of the National Academy of Sciences. 2011; **108**(37):15468-15473

[44] Qin M, Wong A, Seguin D, Gerlai R. Induction of social behavior in zebrafish: Live versus computer animated fish as stimuli. Zebrafish. 2014;**11**(3):185-197

[45] Seibt KJ, Piato AL, da Luz Oliveira R, Capiotti KM, Vianna MR, Bonan CD. Antipsychotic drugs reverse MK-801-induced cognitive and social interaction deficits in Zebrafish (*Danio rerio*). Behavioural Brain Research. 2011;**224**(1):135-139

[46] Parker MO, Annan LV, Kanellopoulos AH, Brock AJ, Combe FJ, Baiamonte M, et al. The utility of Zebrafish to study the mechanisms by which ethanol affects social behavior and anxiety during early brain development. Progress in Neuro-Psychopharmacology & Biological Psychiatry. 2014;**0**:94-100

[47] Bilotta J, Risner ML, Davis EC, Haggbloom SJ. Assessing appetitive choice discrimination learning in zebrafish. Zebrafish. 2005;**2**(4):259-268

[48] Parker MO, Millington ME, Combe FJ, Brennan CH. Development and implementation of a three-choice serial reaction time task for zebrafish (*Danio rerio*). Behavioural Brain Research. 2012;**227**(1):73-80

[49] Levin ED, Chrysanthis E, Yacisin K, Linney E. Chlorpyrifos exposure of developing zebrafish: Effects on survival and long-term effects on response latency and spatial discrimination. Neurotoxicology and Teratology. 2003;**25**(1):51-57

[50] Sison M, Gerlai R. Associative learning in zebrafish (Danio rerio) in the plus maze. Behavioural Brain Research. 2010;**207**(1):99

[51] Colwill RM, Raymond MP, Ferreira L, Escudero H. Visual discrimination learning in zebrafish (*Danio rerio*). Behavioural Processes. 2005;**70**(1):19-31

10

Zebra Fitness: Learning and Anxiety After Physical Exercise in Zebrafish

Mayara Silveira, Jonatas Silveira,
Thais Agues-Barbosa, Miguel Carvalho,
Priscila Silva and Ana Luchiari

Abstract

In the recent years, a new branch of physical training has emerged, the high-intensity interval training (HIIT). In contrast to continued exercise regime used in most of the trainings, HIIT proposes a regime of short periods of maximum intensity exercising and brief less intense recovery periods, which are repeated until complete exhaustion. HIIT is calling the attention of those who search for fast escalation in physical performance; however, the stress caused by this type of training may affect other systems functioning, such as cognition. Thus, we investigated the effects of two physical regime protocols, traditional endurance and HIIT on zebrafish learning, memory, and anxiety-like behav-ior. To that, fish were trained for 30 days and submitted to a latent learning test, objects discrimination test, and novel tank test. Our results showed that HIIT does not affect long lasting memory, evaluated through the latent learning task, but it impairs discriminative learning. On the other hand, both training protocols decrease anxiety-like behavior. This study confirms that zebrafish show good performance in learning tasks and that cognitive performance is dependent upon the regime of physical exercise and cognitive task used.

Keywords: training, latent learning, objects discrimination, novel tank, *Danio rerio*

1. Introduction

It is well known that regular physical activity is an efficient way to improve health in terms of respiratory and cardiovascular functioning [1–3]. In addition to the effects in somatic function-ing, studies with humans and animal models show that physical exercise affects cognition, improving memory, and learning performance [4, 5]. For instance, positive results are reported in terms of attention [6, 7], executive function [8], and motor skills [9].

The benefits of regular training are remarkable in the elderly population. Aging is accompanied by a decline in cognitive abilities, which in some cases occur faster and in a pathological manner [10]. Physical activity retards the cognitive decline, reduces the risk of dementia and neurodegenerative disorders [11–13]. Continuous exercise through lifespan is important to maintain the protective effects in elderly [14] and is also associated with large gray matter in later life [15].

Besides the protective effects against damages due to aging, physical activity can improve cognitive performance in different stages of life. Young adults who practice at least 8 h of physical activity per week present better results in a sustained attention task when compared to individuals with low levels of activity (less than 2 h per week) [16]. In male adolescents, highly intense exercise is associated with the improvement in working memory and in brain-derived neurotrophic factor (BDNF) levels [17]. In children, physical activity improves performance on verbal and mathematical tests, cognitive flexibility and working memory [18, 19].

The benefits mentioned above seem to be related to structural changes in the brain. Physical activity modifies the structure of the hippocampus [20], the most important brain area for learning and memory, and is associated with an increment in gray matter volume probably through BDNF expression [21]. Moreover, it enhances plasticity [22] and stimulates neurogenesis [21].

Another benefit associated with physical activity is the improvement in psychological health. Emotional disorders are common in modern society and impact several domains of one's life. Physical activity has been postulated as a promising treatment strategy [23]. A large body of literature shows that it reduces anxiety, stress, and depression symptoms as it promotes positive mood and greater well-being [23–25]. These behavioral alterations are associated with physical activity modulation of neurotransmitters such as dopamine, associated of the reward system, and serotonin, which plays a role in well-being [26]. Physical activity also interferes in the stress-regulation axis and may buffer stress effects [24].

Many factors can interfere in the beneficial effects of exercise, including individual characteristics such as age, health status, sedentary level, cognitive abilities, and type of exercise [6, 21, 27–30]. More intense activities such as the high-intensity interval training (HIIT) have gained popularity more recently. HIIT is characterized by short periods of maximum intensity exercising interspersed with brief less intense recovery periods. However, it is not clear how this type of exercise impacts health and cognition. HIIT has been elected as the preferred type of training among those who practice physical activity due to its fast conditioning response. However, the trade-off in maintaining HIIT or adopting the traditional training deserves investigation. In this sense, the present study aims at investigating the cognitive and anxiolytic effects of HIIT in relation to traditional training.

Regarding physical activity, zebrafish swimming ability improves with training [31] and enhances associative learning response [32]; therefore, it offers a good organism model to the study of physical activity on cognition.

2. Materials and methods

2.1. Animals and housing

Zebrafish (*Danio rerio*, 3-month age, both sexes) were acquired from a local farm (Natal, Brazil) and maintained in storage tanks (50 L) for a month prior to the experiments. Tanks formed a system with multistage filtration, containing mechanical, biological, and an activated carbon filter, and UV light sterilizing unit. Temperature was kept at $28 \pm 1°C$, and pH and dissolved oxygen were measured regularly. Light was maintained in a 12/12-light/dark cycle. Fish were fed twice a day *ad libitum* with commercial food (38% protein, 4% lipid, Nutricom Pet) and *Artemia salina*. All the experiment procedures were performed with the permission of the Ethical Committee for Animal Use of the Federal University of Rio Grande do Norte (CEUA 054–2016).

2.2. Groups and training

From the stock population, 115 fish were randomly assigned to one of the three groups according to the type of physical training: traditional endurance, HIIT, and sedentary. These fish were housed in separate in smaller tanks (40 × 20 × 25 cm, 15 L) and underwent 7 days of acclimation before beginning the training phase (30 days), after which the tests were performed. The experimental phase was developed in the same environment water quality conditions of the stock fish.

Traditional endurance (n = 34): once a day, fish were transferred to a training tube, always in groups of 5. The training tube consisted in transparent tunnel connected to a 180 L/h submerse water pump (Moto Bomba SARLO S300, 220 V, 60 Hz. Sarlobetter Equipaments Ltda., São Caetano do Sul-SP) placed inside a glass tank [32]. Each group was trained for 30 min, during which the first 15 min fish could not exit the tube and was forced to remain swimming against the water current. Then, the tube was open and fish that could not swim was pushed by the current to the tank. After leaving the tube, they were returned to the home tank until the next session.

HIIT (n = 39): the high-intensity training consisted in a progressive increment of water current intensity. For this, we used water pumps with different powers. During the first 10 days, animals were submitted to a forced swim against the water current powered by a pump of 180 L/H (Moto Bomba SARLO S300, 220 V, 60 Hz. Sarlobetter Equipamentos Ltda., São Caetano do Sul-SP). Between days 11 and 20, a pump with 300 L/h (Moto Bomba SARLO S360) was used and in the last 10 days, it was changed to a 520 L/h pump (Moto Bomba SARLO S520). Training sessions lasted 10 min inside the water current followed by 5 min interval, repeated three times (total time of training: 30 min).

Sedentary (n = 42): fish that was not subjected to training were placed once a day in the training tube without any water current for 30 min. After that, they were returned to the home tank until the next session.

2.3. Latent learning

In this paradigm, fish explore a maze for a training phase without any reinforcement until the test day [33, 34]. The maze has two side arms with the same size (left and right) and a central tunnel that connects the start box to a goal box, where a reward was placed only on the test day. Maze walls were transparent allowing fish to see all the compartments. Exploration/ training sessions occurred once a day, for 30 min, during the last 15 days of physical training. Fish group was placed in the start box for 30 s to habituate and then, the start box's door was opened, allowing fish to explore the maze. From each training group, some animals were trained only to explore the right tunnel of the maze (traditional endurance n = 7; HIIT n = 6; Sedentary n = 7) and some animals were trained to explore only the left tunnel of the maze (traditional endurance n = 9; HIIT n = 8; Sedentary n = 8). The reward chamber was empty during the 15 days of training.

On the test day (31st day of physical training), a small shoal of five fish was positioned in the goal box as reward stimulus to experimental fish. Animals were put individually in the start box and after 30 s the door was lifted, and fish could explore the maze for 10 min. During the test, the right and the left tunnels of the maze were open, so that fish could choose the patch to the goal box. Fish behavior was recorded using a handy cam (Sony Digital Video Camera Recorder; DCR-SX45) above the maze and the following parameters were analyzed: time spent in each tunnel of the maze and in the goal box, first tunnel chosen; latency to enter the shoal area; duration of time with the shoal.

2.4. Objects discrimination test

The object discrimination test consisted in three phases: habituation, memorization, and discrimination phases [35]. The three phases occurred in a 15 L tanks (40 × 20 × 25 cm) with the walls covered with white paper to avoid any extern interference. Habituation phase lasted 5 days (25th–30th days of physical training), during which fish were transferred to the test tank and allowed to explore it for 15 min to reduce possible isolation stress.

On the memorization phase (31st day), two objects (named of A and A') were introduced in the tank test and positioned in each side of the tank, 30 cm from each other. Objects were equal in size, color, and shape. Fish was individually introduced into the tank and explored the objects for 10 min; after which animal was returned to an individual tank. On the next day (32nd day), the discrimination phase occurred, which object A' was replaced by a new one (named object B) with same shape and size but different color. Fish explore the tank and objects for 10 min. During memorization and discrimination phases, fish behavior was recorded from above using a handy cam (Sony Digital Video Camera Recorder; DCR-SX45). Behavior records were analyzed for the time fish spent around each object during both phases [36].

2.5. Novel tank test

The novel tank test is one of the most common tests to evaluate anxiety in zebrafish. It is based on the innate response of fish to dive and freeze when submitted to a new environment/

situation [37]. The test took place after the end of physical training (day 31). Fish was individually transferred to tank (20 × 12 × 15 cm) and behavior was recorded for 10 min with a handy camera (Sony Digital Video Camera Recorder; DCR-SX45) positioned in front of the tank. It evaluated average swimming velocity, total distance traveled, freezing duration, and distance from the bottom of the tank.

2.6. Statistical analysis

For the latent learning test, one-way analysis of variance (ANOVA) was performed to investigate the main effect of training regime and behavioral parameters during the test trial. When significant effects were identified, post hoc Student Newman Keuls test was conducted to reveal significant ($p \leq 0.05$) group differences.

We compared the objects exploration time in memorization and discrimination phases, and also between the two phases using Two-Way ANOVA followed by Student Newman Keuls when significance was found.

We also calculated an index of how much fish explored each object during both phases. The exploration index for the memorization phase (exploration of memorization = E_m = A1 + A2) and discrimination index for the discrimination phase (discrimination index = D_i = B − A3) were calculated to verify whether there were differences in object exploration time between groups [38]. Correlations between exploration in the memorization and discrimination phases was compared using Simple Linear Regression (Pearson's correlation).

All the locomotor parameters from the novel tank test were statistically compared using One-Way ANOVA. For all tests, we considered the probability level of $p < 0.05$ for statistical significance.

3. Results

3.1. Latent learning

Figure 1 presents the number of correct choice fish made during the test day and the time spent in each tunnel on the test day, when both the right and the left tunnels of the maze were open. One-Way ANOVA showed neither difference in the number of correct choice between the three groups ($F(2,44) = 0.76$, $p = 0.47$; **Figure 1a**) nor in the number of correct choice considering the tunnel where fish was trained before (right trained fish: $F(2,20) = 0.61$, $p = 0.55$; left trained fish: $F(2,24) = 0.32$, $p = 0.72$; **Figure 1b**). Also, time spent in the correct tunnel did not differ between groups ($F(2,44) = 0.04$, $p = 0.96$; **Figure 1c**) and did not differ considering the tunnel where fish was trained before (right trained fish: $F(2,20) = 0.10$, $p = 0.90$; left trained fish: $F(2,24) = 0.31$, $p = 0.74$; **Figure 1d**).

However, during the test day, in which a shoal was presented at the reward area, animals from HIIT group showed lower latency to enter in shoal area compared to the others groups

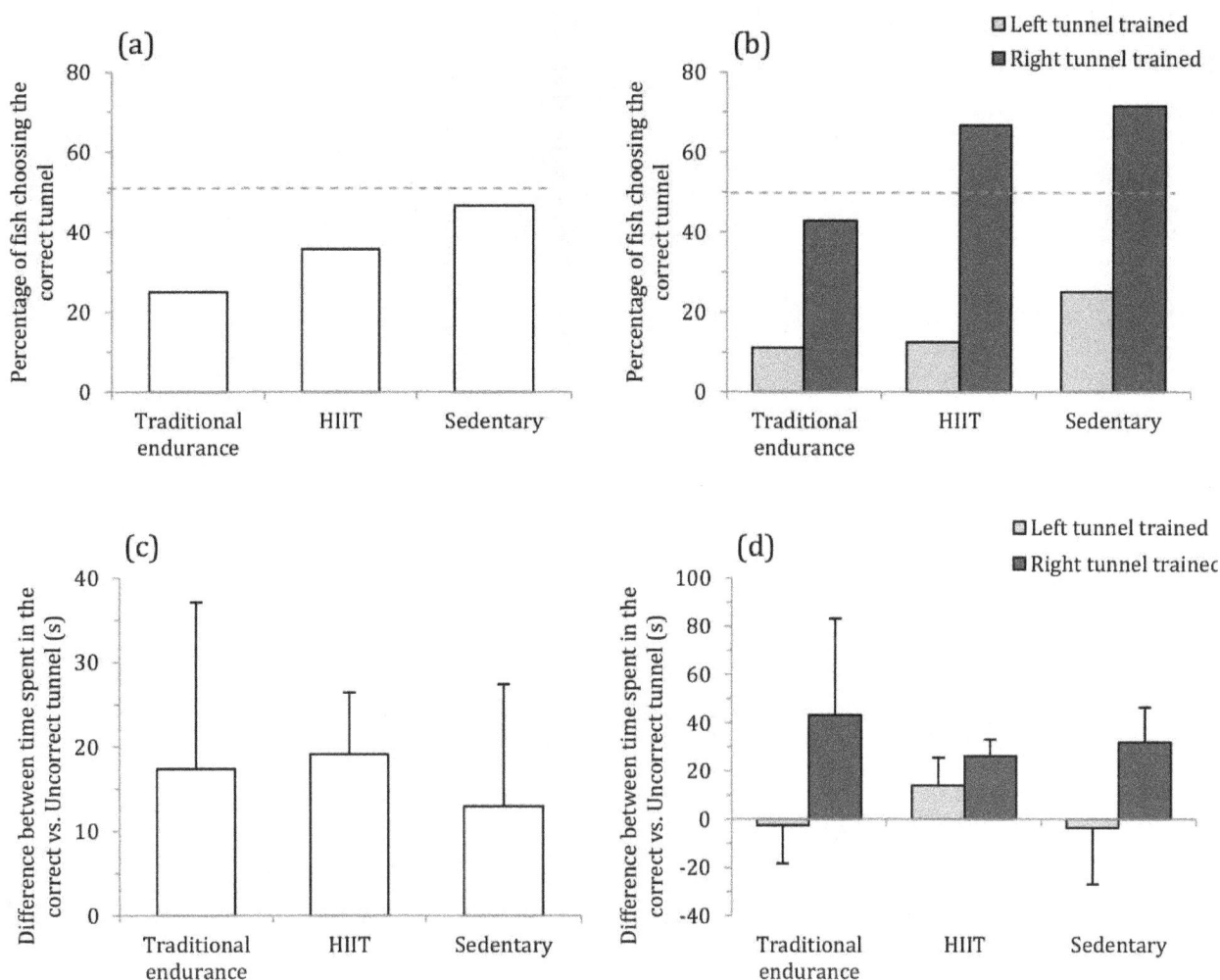

Figure 1. Number of fish that chose the correct tunnel (a, b) and time spent in each tunnel (c, d) after leaving the start box on the probe trial, during which both the right and the left tunnels of the maze were open. The correct choice was the left tunnel for fish that were previously trained with the left tunnel open, and it was the right tunnel for fish that were trained with the right tunnel open. Panel (a) and (c) show the total number of fish that made correct choice and time spent in the correct tunnel irrespective of left or right tunnel training for each treatment. Panel (b) and (d) show the correct choice and time spent in the correct tunnel of each group considering whether the fish were trained with the left or right tunnel: light gray bars represent the fish trained with the left tunnel open and dark bars represent the fish trained with the right tunnel open. Training regimes were: Traditional endurance (n = 16), HIIT (n = 14) and sedentary (n = 15). The physical training conditions are shown on the x-axis: traditional endurance group was trained daily in a current water tube until fish could not swim against it and was pushed to the calm water; high-intensity interval training (HIIT) group was trained to maximum swim against the current three times of 10 min with 5 min interval; sedentary group was put inside the training tube for the same period the other groups trained, but no water current was generated. All panels show that there were no differences between groups. For further details of the results of statistical analysis, see results.

(One-Way ANOVA $F(2,44) = 3.44$, $p = 0.04$), and the same HIIT group spent less time in shoal area (One-Way ANOVA $F(2,44) = 20.70$, $p < 0.001$), as shown in **Figure 2**.

3.2. Objects discrimination

Figure 3 depicts the difference between the time exploring objects A and A' in the memorization phase and A and B in the discrimination phase for all the training groups. For traditional

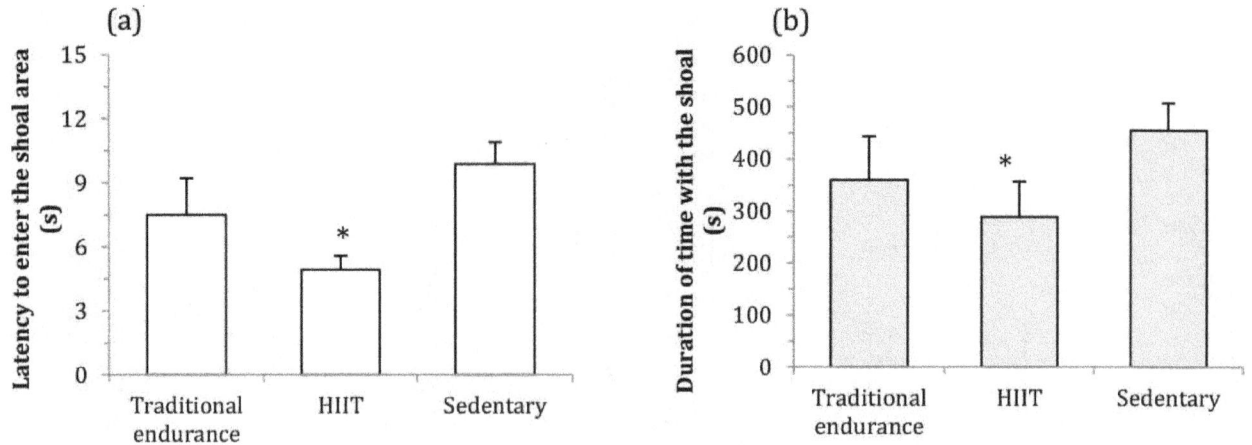

Figure 2. (a) Latency to leave the start box and (b) duration of time zebrafish spent in the goal box during the probe trial. Bars show mean ± SEM; sample size equals 16 for traditional endurance group, 14 for HIIT group, and 15 for sedentary group. The physical training conditions are shown on the x-axis: traditional endurance group was trained daily in a current water tube until fish could not swim against it and was pushed to the calm water; high-intensity interval training (HIIT) group was trained to maximum swim against the current three times of 10 min with 5 min interval; sedentary group was put inside the training tube for the same period the other groups trained, but no water current was generated. During the latent learning training sections, fish were allowed to explore the maze with only one tunnel open. During the probe trial, both tunnels were open and fish were allowed to use the tunnels to reach the goal box, where a conspecific shoal was presented. Asterisk indicates statistical difference between groups (one-way ANOVA, $p < 0.05$).

endurance group, Two-Way ANOVA of this data set revealed a significant effect of the objects ($F_{(1,56)} = 4.02$, $p = 0.04$), no effects of the phase ($F_{(1,56)} = 1.63$, $p = 0.21$), but the objects × phase interaction ($F_{(1,56)} = 9.01$, $p = 0.006$) was also significant. Student Newman Keuls test showed that exploration time of the novel object on the discrimination phase significantly ($p < 0.01$) differed from exploration time of the other objects (**Figure 3a**). For the HIIT group, Two-Way ANOVA showed no significant effect of the objects ($F_{(1,56)} = 0.20$, $p = 0.65$), no effects of the phase ($F_{(1,56)} = 1.85$, $p = 0.18$), and also no effects of interaction ($F_{(1,56)} = 0.50$, $p = 0.48$) (**Figure 3b**). For the sedentary group, Two-Way ANOVA indicates significant effect of the objects ($F_{(1,56)} = 29.93$, $p = 0.001$), no effects of the phase ($F_{(1,56)} = 0.78$, $p = 0.38$), but again significance was observed for the objects × phase interaction ($F_{(1,56)} = 9.78$, $p = 0.003$). Student Newman Keuls test revealed that exploration time of the novel object on the discrimination phase was significant different ($p < 0.01$) from exploration time of the other objects (**Figure 3c**).

Figure 4 shows the correlation between exploration index (E_m; memorization phase) and discrimination index (D_i; discrimination phase) for the training groups. For the traditional endurance group, E_m and D_i were significant (ANOVA, $F_{(1,14)} = 5.67$, $p = 0.03$), showing a directly proportional relationship, as demonstrated by the angular coefficient ($y = 0.66x - 1.90$) and Pearson's correlation coefficient ($r = 0.04$; **Figure 4a**). For HIIT, there was no significant correlation between E_m and D_i (Pearson's correlation coefficient $r = 0.00007$, $p = 0.39$; **Figure 4b**). For the sedentary group, E_m and D_i were significant (ANOVA, $F_{(1,7)} = 3.67$, $p = 0.04$) and showed proportional relationship, as observed from the angular coefficient ($y = 0.42x + 28.81$) and Pearson's correlation coefficient ($r = 0.10$; **Figure 4c**).

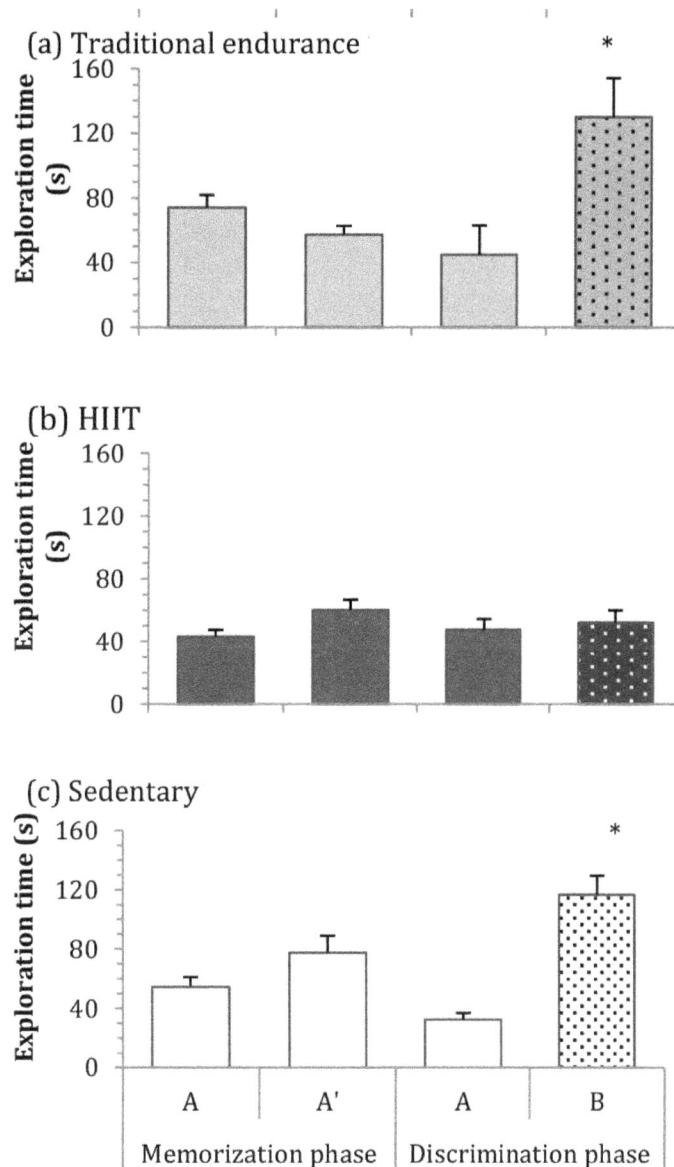

Figure 3. Zebrafish exploration time for objects A vs. A' (memorization phase), or A vs. B (discrimination phase) for the physical training exposure regime groups: (a) traditional endurance, in which fish were trained daily in a current water tube until it could not swim against it and was pushed to the calm water (n = 8), (b) high-intensity interval training (HIIT) group, in which fish were trained to maximum swim against the current 3 times of 10 min with 5 min interval (n = 14), and (c) sedentary group, in which fish were put inside the training tube for the same period the other groups trained, but no water current was generated (n = 15). Bars mean exploration time + SEM in each object, in the memorization phase (two equal objects) and in the discrimination phase (two different objects). Fish were observed for 10 min in each section. Asterisk indicates statistical difference between fish objects exploration (two-way ANOVA, p < 0.05).

3.3. Novel tank test

Figure 5 presents difference between the training groups related to average speed, total distance traveled, freezing behavior and distance from the bottom of the tank. One-Way ANOVA showed that training regime did not affect average speed (F(2,33) = 2.07, p = 0.14; **Figure 5a**), total distance traveled (F(2,33) = 1.89, p = 0.17; **Figure 5b**), and distance from the bottom of

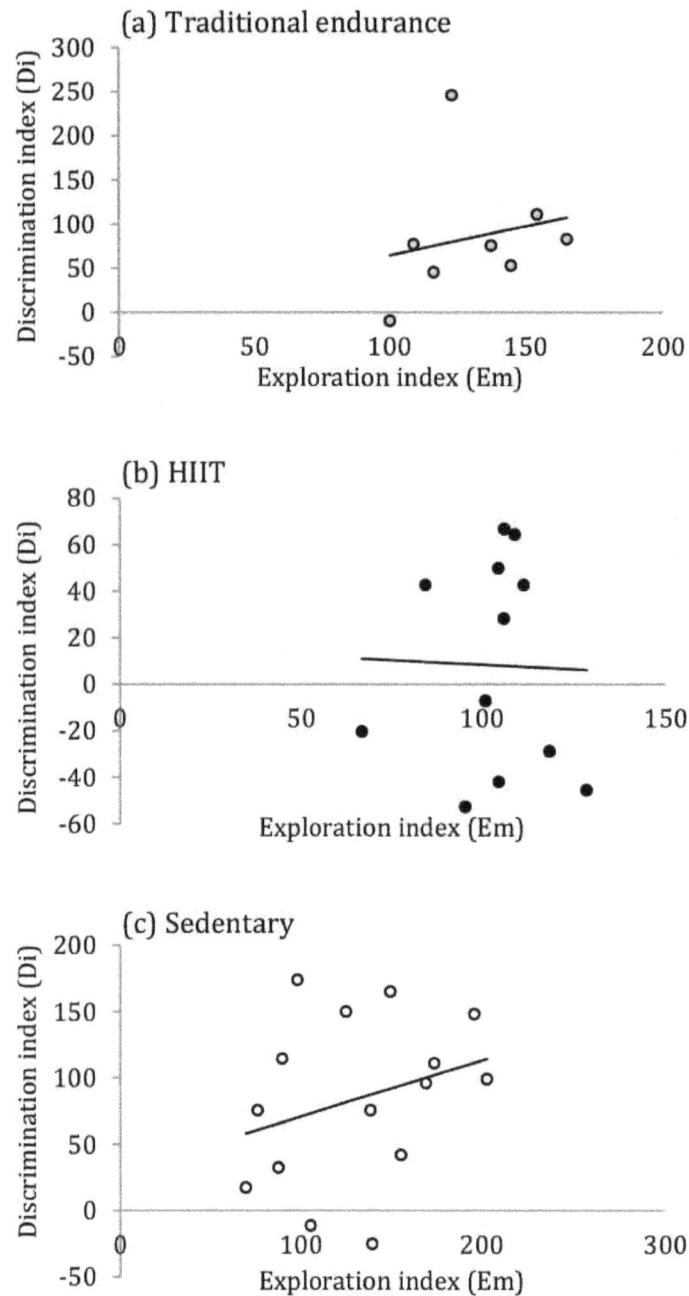

Figure 4. Linear regression between discrimination index (D_i) and exploration index (E_m) in the memorization phase. (a) Traditional endurance, in which fish were trained daily in a current water tube until it could not swim against it and was pushed to the calm water (n = 8), (b) high-intensity interval training (HIIT) group, in which fish were trained to maximum swim against the current three times of 10 min with 5 min interval (n = 14), and (c) sedentary group, in which fish were put inside the training tube for the same period the other groups trained, but no water current was generated (n = 15). For further details of the results of statistical analysis, see results.

the tank ($F_{(2,33)} = 0.08$, p = 0.92; **Figure 5d**). However, freezing behavior, a characteristic mainly related to anxiety, was affected by exercising. One-Way ANOVA showed that sedentary group presented higher freezing behavior than the traditional endurance and HIIT groups ($F_{(2,33)} = 9.62$, p < 0.001; **Figure 5c**).

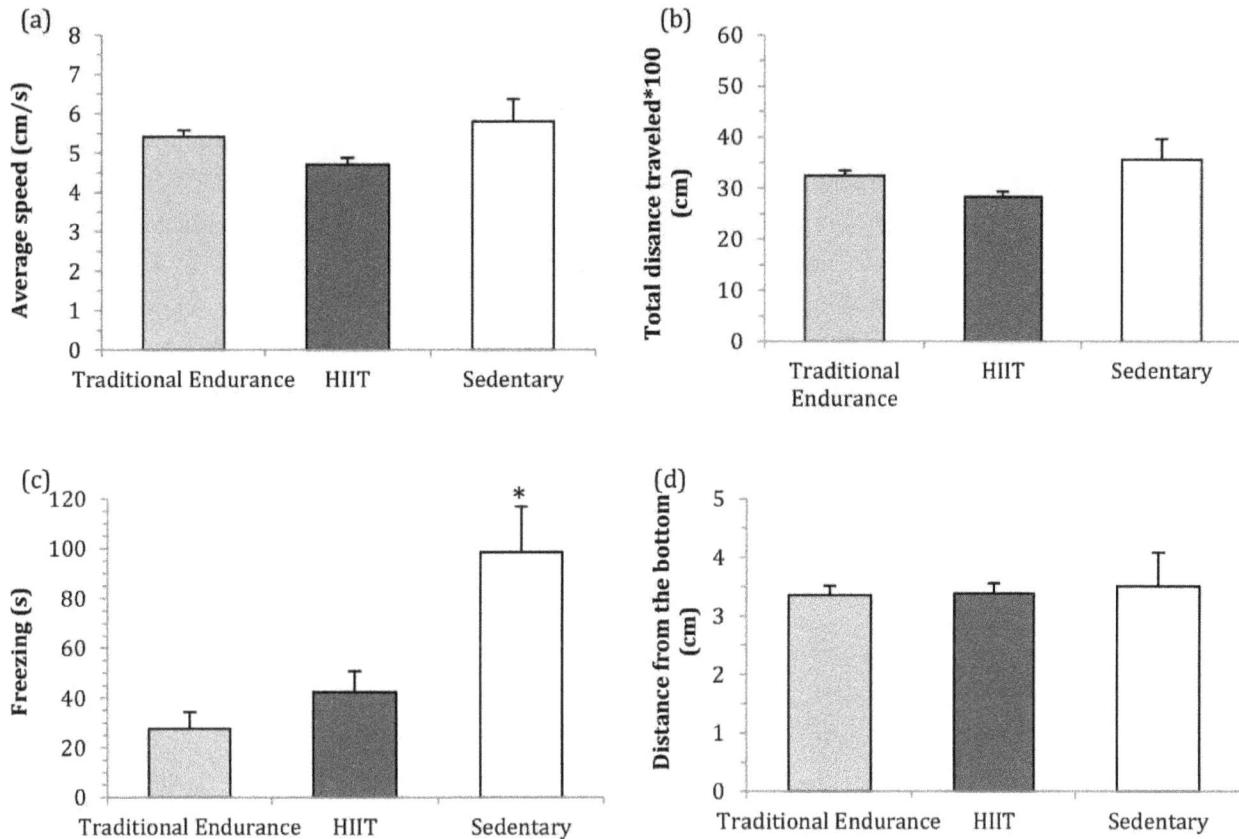

Figure 5. Behavioral analysis in the novel tank test for each treatment: traditional endurance (n = 10), HIIT (n = 11), and sedentary (n = 12). Behavioral tracking applied to compare (a) average speed swimming, (b) total distance traveled by the fish, (c) freezing behavior, and (d) distance from the bottom of the tank between the three groups. Error bars represent standard error. Data corresponds to 15 min of behavioral observation during the test. Asterisk indicates statistically differences (one-way ANOVA, p < 0.05).

4. Discussion

In this study, we show the effects of traditional endurance and high-intensity interval training (HIIT) on the learning and anxiety-like behavior in adult zebrafish. Our results evidence that HIIT does not affect long lasting memory, as observed in the latent learning task, but it impairs discriminative learning, as observed in the object discrimination task. We reinforce the zebrafish intrinsic explorative behavior, both directed to novelty and social groups, and showed that physical training affects these responses. However, both HIIT and traditional exercise regime seems to have decreased anxiety-like behavior in zebrafish. This study confirms that zebrafish show good performance in learning tasks and that cognitive performance is dependent upon the regime of physical exercise and cognitive task used.

Our results are in agreement with other studies on learning in zebrafish [33–35, 38, 39]. Latent learning has been tested in zebrafish previously [33, 34] and the results are in line with those reported here. Even the lateralization (right tunnel bias in the first choice) was observed in these other studies using the same protocol we used. Regarding this response (right turn), the mechanisms that govern this bias are still not understood but a growing body of literature has pointed out lateralization in zebrafish brain [33, 34, 40–42] probably due to asymmetrical biochemical and neurobiological processes. Regarding the objects discrimination using the same protocol we have presented, Oliveira et al. [35], Santos et al. [39], and Pinheiro-da-Silva et al. [38] showed zebrafish ability to recognize the novel object in the discrimination phase of the tests. However, even though the zebrafish is considered a translational model for humans, physical exercise has received little attention, and as far as we know, only two studies approach this issue. One of them relates physical exercise and aging in a protocol that emphasizes swimming performance [31], while the other evaluates the effects of traditional training and associative learning [32]. In the former, the authors have shown that training enhances cognitive performance in zebrafish compared to sedentary fish. However, in the present study, we use both traditional endurance and HIIT to compare how these two different types of training affect cognition. We observed that chronic traditional training (i.e. daily walking, running, or cycling) allowed zebrafish to latent learning and object discrimination, what was not observed in high-intensity training.

Many recent studies have been approaching the high-intensity exercise protocols [2, 8, 28, 43]. It is of particular interest because this mode of training has spread through most of the gyms and young people have been adopting mainly this type of training due to the fast conditioning response it promotes [8]. In the present study, we intent to simulate HIIT protocol by exercising the fish acutely in a high-intensity swimming activity and then letting them rest for a short time before repeating the process. This procedure has negatively affected the fish ability to discriminate objects (**Figures 3** and **4**) but is had no effects on latent learning (**Figure 1**). We also observed that HIIT fish decreased time to reach the shoal in the latent learning test, suggesting it performed the task faster than the other groups. However, HIIT fish showed decreased time with the shoal. Shoaling is a behavioral pattern usually observed in zebrafish due to its high social behavior [44, 45], suggesting the training protocol decreased fish interest in maintaining social interaction. We believe that HIIT may have caused an increased exploratory effect, leading the fish to explore the tank more and shoaling less in the latent learning test. The same effect may explain why fish did not discriminate objects: HIIT may have made the fish to explore the tank of the objects more than focus its attention on the objects itself. These results seem to agree with the decreased anxiety-like behavior (decreased freezing) observed in the novel tank test, which may lead fish to be more explorative than presenting the need for social interaction.

As it is known, many variables may affect exercise training on cognition, such as the cognitive function examined and the level of physical training [2]. Regular physical exercise promotes a number of benefits in the organisms [46, 47] and may have positive effects on cognitive function [32, 48]. In the present study, we confirm that traditional training protocols (continued exercising without high-intensity periods) had beneficial consequences: fish behaves as the control (sedentary) in the latent learning and in the objects discrimination task, but it decreased anxiety-like behavior in the novel tank. Thus, while it seems that traditional endurance cognitive function is comparable to sedentary fish, the anxiety response was significant lower in the fish continuous trained.

Perhaps the most interesting effect, we found in this study was the decreased freezing behavior in the physical training groups. These fish were exposed to training for a prolonged period of time, and tested to the novel tank only at the end of the training period. Different from the cognitive tasks, in which fish were allowed to explore the tank in advance, in the novel tank protocol, fish had only one chance to explore and learn about the tank. Somehow fish used to physical training perceived the novel tank as less threatening, and spent its time exploring the tank instead of showing freezing behavior. In this sense, physical exercise improves one's performance when facing a novel situation, response that is lost when the animal is re-submitted to the tank several times to reduce stress, as we have done on the two cognitive tasks used in the present study.

In fact, our results reinforce the zebrafish as a valuable model organism for throughput screening of behavioral- and cognitive-related training regimes. While physical exercise is highly encouraged and indeed recommended due to several health benefits related to its chronic practice, it seems that the new category of high-intensity training still needs investigations and search for the balance between the benefits of exercising and the damages of acute intense stress. In this sense, our study is far from responding the question: other cognitive protocols should be tested, novel HIIT protocols need to be developed, and several physiological and neural parameters need to be measured, such as O_2 consumption, metabolic rate, catecholamine and corticosteroid release indices, and brain-derived neurotrophic factor levels. Moreover, an alternative form of high-intensity training that also demands attention, coordination, and decision making should also be approached in future studies, such as the game-based activities. Some authors have studied these activities in humans and showed positive cognitive results afterwards [49, 50].

Therefore, while our results are robust in showing there are several effects of the physical training regime on behavior and cognition, we still need to understand how exercising models the body and the brain to cause such effects and what is the extent to which exercising if indeed beneficial. Taken together, these results indicate that high-intensity exercising implies in decreased short-term learning and confirm the zebrafish as a trustful, reliable, and efficient model for basic translational research of the effects of physical activity/training on cognition and behavior.

5. Conclusions

Overall, our results zebrafish is a profitable animal model for basic translational research. Physical exercise in the modern society is medically recommended and rigorously followed in search of increased health and willingness to study/work. However, not all exercising regimes are adequate to anyone and the effects on the body and the brain may exceed the beneficial limit and cause injury, as the decline in short-term learning and spatial navigation observed herein after high-intensity training regime. Therefore, while exercising is important for health, it seems that a trade-off between exercise type and intensity should be taken into account when searching for the enhancement of body and brain. Still, additional studies are needed for a thorough understanding of the effects of physical training and the limits it may impose to our health.

Acknowledgements

We thank Ms. Peripolli C. and for help in collecting data for this article.

Conflict of interest

Hereby we confirm the absence of any conflict of interest related to this work.

Author details

Mayara Silveira, Jonatas Silveira, Thais Agues-Barbosa, Miguel Carvalho, Priscila Silva and Ana Luchiari*

*Address all correspondence to: analuchiari@yahoo.com.br

Physiology Department, Biosciences Center, Federal University of Rio Grande do Norte, Rio Grande do Norte, Brazil

References

[1] Antunes HKM, Santos RF, R. Cassilhas, R.V.T. Santos, Bueno OFA, De Mello MT. Exercício físico e função cognitiva: Uma revisão. Revista Brasileira de Medicina do Esporte. 2006;**12**:108-114. DOI: 10.1590/S1517-86922006000200011

[2] Cooper, Simon B, Dring, Karah J. High-intensity intermittent exercise: Effect on young people's cardiometabolic health and cognition, Current Sports Medicine Reports. 2016;**15**:245-251. DOI: http://dx.doi.org/10.1249/JSR.0000000000000273

[3] Warburton DER, Nicol CW, Bredin SSD. Health benefits of physical activity: The evidence. Canadian Medical Association Journal. 2006;**174**:801-809

[4] Pontifex MB, Gwizdala KL, Parks AC, Pfeiffer KA, Fenn KM. The association between physical activity during the day and long-term memory stability. Scientific Reports. 2016; **6**:1-9. DOI: 10.1038/srep38148

[5] Liu F, Sulpizio S, Kornpetpanee S, Job R. It takes biking to learn: Physical activity improves learning a second language. PLoS One. 2017;**12**:1-15. DOI: 10.1371/journal. pone.0177624

[6] Griffin ÉW, Mullally S, Foley C, Warmington SA, Mara SMO', Kelly ÁM. Aerobic exercise improves hippocampal function and increases BDNF in the serum of young adult males. Physiology & Behavior. 2011;**104**:934-941. DOI: 10.1016/j.physbeh.2011.06.005

[7] Luque-Casado A, Perakakis P, Hillman CH, Kao SC, Llorens F, Guerra P, Sanabria D. Differences in sustained attention capacity as a function of aerobic fitness. Medicine and Science in Sports and Exercise. 2016;**48**:887-895. DOI: 10.1249/MSS.0000000000000857

[8] Moreau D, Kirk IJ, Waldie KE. High-intensity training enhances executive function in children in a randomized, placebo-controlled trial. eLife. 2017;**6**:1-26. DOI: 10.7554/eLife.25062

[9] Burns RD, Fu Y, Fang Y, Hannon JC, Brusseau TA. Effect of a 12-week physical activity program on gross motor skills in children. Perceptual and Motor Skills. 2017;**124**:1121-1133

[10] Wong RY. Physical exercise, cognition, and function in older people. Journal of the American Medical Directors Association. 2017;**18**:282-283. DOI: 10.1016/j.jamda.2017.01.002

[11] Abbott RD, White LR, Ross GW, Masaki KH, Curb JD, Petrovitch H. Walking and dementia in physically capable elderly men. Journal of the American Medical Association. 2004;**292**:1447-1453

[12] Barnes DE, Yaffe K. The projected effect of risk factor reduction on Alzheimer's disease prevalence. Lancet Neurology. 2011;**10**:819-828

[13] Pignatti F, Rozzini R, Trabucchi M. Physical activity and cognitive decline in elderly persons. Archives of Internal Medicine. 2002;**162**:361. https://jamanetwork.com/journals/jamainternalmedicine/fullarticle/210935

[14] Richards M, Hardy R, Wadsworth MEJ. Does active leisure protect cognition? Evidence from a national birth cohort. Social Science & Medicine. 2003;**56**:785-792

[15] Rovio S, Spulber G, Nieminen LJ, Niskanen E, Winblad B, Tuomilehto J, Nissinen A, Soininen H, Kivipelto M. The effect of midlife physical activity on structural brain changes in the elderly. Neurobiology of Aging. 2010;**31**:1927-1936

[16] Luque-Casado A, Perakakis P, Ciria LF, Sanabria D. Transient autonomic responses during sustained attention in high and low fit young adults. Scientific Reports. 2016;**6**:27556. DOI: 10.1038/srep27556

[17] Jeon YK, Ha CH. The effect of exercise intensity on brain derived neurotrophic factor and memory in adolescents. Environmental Health and Preventive Medicine. 2017;**22**:27. DOI: 10.1186/s12199-017-0643-6

[18] Sibley BA, Etnier JL. The relationship between physical activity and cognition in children: A meta-analysis. Pediatric Exercise Science. 2003;**15**:243-256

[19] Ludyga S, Gerber M, Brand S, Holsboer-Trachsler E, Pühse U. Acute effects of moderate aerobic exercise on specific aspects of executive function in different age and fitness groups: A meta-analysis. Psychophysiology. 2016;**11**:1611-1626. DOI: 10.1111/psyp.12736

[20] Makizako H, Liu-Ambrose T, Shimada H, Doi T, Park H, Tsutsumimoto K, Uemura K, Suzuki T. Moderate-intensity physical activity, hippocampal volume, and memory in

older adults with mild cognitive impairment. The Journals of Gerontology. Series A, Biological Sciences and Medical Sciences. 2014;**70**:480-486

[21] Erickson KI, Hillman CH, Kramer AF. Physical activity, brain, and cognition. Current Opinion in Behavioral Sciences. 2015;**4**:27-32. DOI: 10.1016/j.cobeha.2015.01.005

[22] Cotman CW, Berchtold NC. Exercise: A behavioral intervention to enhance brain health and plasticity. Trends in Neurosciences. 2002;**25**:295-301

[23] Morgan JA, Singhal G, Corrigan F, Jaehne EJ, Jawahar MC, Baune BT. The effects of aerobic exercise on depression-like, anxiety-like, and cognition-like behaviours over the healthy adult lifespan of C57BL/6 mice. Behavioural Brain Research. 2018;**337**:193-203. DOI: 10.1016/j.bbr.2017.09.022

[24] Strasser B, Fuchs D. Role of physical activity and diet on mood, behavior, and cognition. Neurology Psychiatry and Brain Research. 2015;**21**:118-126. DOI: 10.1016/j.npbr. 2015.07.002

[25] Greer TL, Furman JL, Trivedi MH. Evaluation of the benefits of exercise on cognition in major depressive disorder. General Hospital Psychiatry. 2017:1-7. DOI: 10.1016/j. genhosppsych.2017.06.002

[26] Lin T-W, Kuo Y-M. Exercise benefits brain function: The monoamine connection. Brain Sciences. 2013;**3**:39-53

[27] Loprinzi PD, Edwards MK, Frith E. Potential avenues for exercise to activate episodic memory-related pathways: A narrative review. The European Journal of Neuroscience. 2017;**46**:2067-2077. DOI: 10.1111/ejn.13644

[28] Hwang J, Brothers RM, Castelli DM, Glowacki EM, Chen YT, Salinas MM, Kim J, Jung Y, Calvert HG. Acute high-intensity exercise-induced cognitive enhancement and brain-derived neurotrophic factor in young, healthy adults. Neuroscience Letters. 2016; **630**:247-253

[29] Heisz JJ, Vandermorris S, Wu J, McIntosh AR, Ryan JD. Age differences in the association of physical activity, sociocognitive engagement, and TV viewing on face memory. Health Psychology. 2015;**34**:83

[30] Ruscheweyh R, Willemer C, Krüger K, Duning T, Warnecke T, Sommer J, Völker K, Ho HV, Mooren F, Knecht S. Physical activity and memory functions: An interventional study. Neurobiology of Aging. 2011;**32**:1304-1319

[31] Gilbert MJH, Zerulla TC, Tierney KB. Zebrafish (Danio Rerio) as a model for the study of aging and exercise: Physical ability and trainability decrease with age. Experimental Gerontology. 2013;**50**:106-113. DOI: 10.1016/j.exger.2013.11.013

[32] Luchiari AC, Chacon DMM. Physical exercise improves learning in zebrafish, Danio Rerio. Behavioural Processes. 2013;**100**:44-47. DOI: 10.1016/j.beproc.2013.07.020

[33] Gómez-Laplaza LM, Gerlai R. Latent learning in zebrafish (Danio Rerio). Behavioural Brain Research. 2010;**208**:509-515

[34] Luchiari AC, Salajan DC, Gerlai R. Acute and chronic alcohol administration: Effects on performance of zebrafish in a latent learning task. Behavioural Brain Research. 2015;**282**:76-83. DOI: 10.1016/j.bbr.2014.12.013

[35] Oliveira J, Silveira M, Chacon D, Luchiari A. The zebrafish world of colors and shapes: Preference and discrimination. Zebrafish. 2015;**0**:150212075251004. DOI: 10.1089/zeb.2014.1019

[36] Lucon-Xiccato T, Dadda M. Assessing memory in zebrafish using the one-trial test. Behavioural Processes. 2014;**106**:1-4

[37] Meshalkina DA, Kizlyk MN, Kysil EV, Collier AD, Echevarria DJ, Abreu MS, Barcellos LJG, Song C, Kalueff AV. Understanding zebrafish cognition. Behavioural Processes. 2017;**141**:229-241. DOI: 10.1016/j.beproc.2016.11.020

[38] Pinheiro-da-Silva J, Silva PF, Nogueira MB, Luchiari AC. Sleep deprivation effects on object discrimination task in zebrafish (Danio Rerio). Animal Cognition. 2016:1-11. DOI: 10.1007/s10071-016-1034-x

[39] Santos LC, Ruiz-Oliveira J, Oliveira JJ, Silva PF, Luchiari AC. Irish coffee: Effects of alcohol and caffeine on object discrimination in zebrafish. Pharmacology, Biochemistry, and Behavior. 2016;**143**:34-43. DOI: 10.1016/j.pbb.2016.01.013

[40] Miklosi A, Andrew RJ. Right eye use associated with decision to bite in zebrafish. Behavioural Brain Research. 1999;**105**:199-205

[41] Miklósi Á, Andrew RJ, Gasparini S. Role of right hemifield in visual control of approach to target in zebrafish. Behavioural Brain Research. 2001;**122**:57-65

[42] Barth KA, Miklosi A, Watkins J, Bianco IH, Wilson SW, Andrew RJ. Fsi zebrafish show concordant reversal of laterality of viscera, neuroanatomy, and a subset of behavioral responses. Current Biology. 2005;**15**:844-850

[43] Villelabeitia-Jaureguizar K, Vicente-Campos D, Senen AB, Jiménez VH, Garrido-Lestache MEB, Chicharro JL. Effects of high-intensity interval versus continuous exercise training on post-exercise heart rate recovery in coronary heart-disease patients. International Journal of Cardiology. 2017;**244**:17-23. DOI: 10.1016/j.ijcard.2017.06.067

[44] Engeszer RE, Patterson LB, Rao AA, Parichy DM. Zebrafish in the wild: A review of natural history and new notes from the field. Zebrafish. 2007;**4**:21-40. DOI: 10.1089/zeb.2006.9997

[45] Dreosti E, Lopes G, Kampff AR, Wilson SW. Development of social behavior in young zebrafish. Frontiers in Neural Circuits. 2015;**9**

[46] Zierath JR, Wallberg-Henriksson H. Looking ahead perspective: Where will the future of exercise biology take us? Cell Metabolism. 2015;**22**:25-30. DOI: 10.1016/j.cmet.2015.06.015

[47] Hawley JA, Hargreaves M, Joyner MJ, Zierath JR. Integrative biology of exercise. Cell. 2014;**159**:738-749. DOI: 10.1016/j.cell.2014.10.029

[48] Fernandes J, Arida RM, Gomez-Pinilla F. Physical exercise as an epigenetic modulator of brain plasticity and cognition. Neuroscience and Biobehavioral Reviews. 2017;**80**:443-456. DOI: 10.1016/j.neubiorev.2017.06.012

[49] Budde H, Voelcker-Rehage C, Pietraßyk-Kendziorra S, Ribeiro P, Tidow G. Acute coordinative exercise improves attentional performance in adolescents. Neuroscience Letters. 2008;**441**:219-223

[50] Pesce C. An integrated approach to the effect of acute and chronic exercise on cognition: The linked role of individual and task constraints. In: McMorris T, Tomporowski PD, Audiffren M, editors. Exercise and Cognitive Function. New Jersey: John Wiley & Sons; 2009. p. 213-226

11

Transient-Receptor Potential (TRP) and Acid-Sensing Ion Channels (ASICs) in the Sensory Organs of Adult Zebrafish

Antonino Germanà, Juan D. Muriel, Ramón Cobo,
Olivia García-Suárez, Juan Cobo and José A. Vega

Abstract

Sensory information from the aquatic environment is required for life and survival of zebrafish. Changes in the environment are detected by specialized sensory cells that convert different types of stimuli into electric energy, thus originating an organ-specific transduction. Ion channels are at the basis of each sensory modality and are responsible or are required for detecting thermal, chemical, or mechanical stimuli but also for more complex sensory processes as hearing, olfaction, taste, or vision. The capacity of the sensory cells to preferentially detect a specific stimulus is the result of a characteristic combination of different ion channels. This chapter summarizes the current knowledge about the occurrence and localization of ion channels in sensory organs of zebrafish belonging to the superfamilies of transient-receptor potential and acid-sensing ion channels that are involved in different qualities of sensibility superfamilies in the sensory organs of zebrafish. This animal model is currently used to study some human pathologies in which ion channels are involved. Furthermore, zebrafish is regarded as an ideal model to study in vivo the transient-receptor potential ion channels.

Keywords: sensory organs, sensibility, transient-receptor potential ion channels, acid-sensing ion channels, zebrafish

1. Introduction

Sensory information from the environment is required for life and survival, and it is detected by specialized cells which together make up the sensory system. In fishes the sensory system consists of specialized sensory organs (SO) able to detect light, mechanical and chemical environmental stimuli [1]. SO contain differentiated and specialized sensory cells that convert different types of stimuli into electric energy, thus originating an organ-specific transduction. Actually it is accepted that ion channels are at the basis of each sensory modality and are responsible or are required for detecting thermal, chemical, or mechanical stimuli [2–4].

The identification of ion channels selectively activated by specific stimuli supported the concept that the expression of a particular ion channel confers selectivity to respond to a unique stimulus. Nevertheless, the ion channels originally proposed as specific transducers are not selectively associated with the distinct types of sensibility. In fact, it has been observed that ion channels originally associated with one particular stimulus can be activated by different stimuli and are expressed in sensory cells functionally specific for other sensitivities. In other words, a specific ion channel can be expressed in more than a sensory cell type, and each cell type may express more than one type of ion channel. Thus, the capacity of the sensory cells of a SO to preferentially detect a specific stimulus is the result of a characteristic combination of different ion channels [5–7]. On the other hand, to be a reasonable candidate for sensing and/or transducing a stimulus, an ion channel must be expressed in the right place. Thus, the sensory cells of SO are thought to express ion channels that can act as sensors/transducers of the sensibility they are deputy. For example, some acid-sensing ion channels (ASICs) are presumably involved in mechanosensation and therefore are expressed in the mechanoreceptor cells. Also, in detecting chemical properties of food, ASICs participate, and consistently they are present in sensory cells of taste buds.

The interest for the presence of ion channels in the zebrafish SO is because it is a model to study some human pathologies in which ion channels are involved, related to vision [8–10], hearing and balance [11–13], taste [14], or olfaction [15]. Furthermore, zebrafish is regarded as an ideal model to study in vivo the transient-receptor potential (TRP) ion channels [16, 17].

Thus, this chapter summarizes the current knowledge about the occurrence and localization in SO of zebrafish of ion channels belonging to the superfamilies of TRP and ASICs that are involved in different qualities of sensibility.

2. The superfamily of transient-receptor potential (TRP) ion channels

TRPs are nonselective cation channels, of which few are highly Ca^{2+} selective and some are permeable for highly hydrated Mg^{2+}. The TRP superfamily is subdivided into seven subfamilies: TRPC (canonical), TRPV (vanilloid), TRPM (melastatin), TRPP (polycystin), TRPML (mucolipin), TRPA (ankyrin), and TRPN (NOMPC-like); the latter one is found only in invertebrates and fishes [18]. In yeast, the eighth TRP family was recently identified and named as TRPY [19].

At least 28 different TRP proteins have been identified in mammals. Structurally, a typical TRP protein contains six putative transmembrane domains (S1–S6) with a pore-forming reentrant loop between S5 and S6. Intracellular N- and C-termini are variable in length and consist of a variety of domains [3, 20, 21]. This ion channel superfamily shows a variety of gating mechanisms with modes of activation ranging from ligand binding, voltage, and changes in temperature to covalent modifications of nucleophilic residues (see for a review [3, 22, 23]).

TRP channels serve diverse functions. The members of the TRP superfamily with potential capability for mediating mechanosensing include TRPC1, TRPC3, TRPC5, TRPC6, TRPV4, TRPM3, TRPM7, TRPP1, and TRPP2 [3, 23–27]. Some of them have been suggested as candidates for the mechanotransduction channel in the inner ear vertebrate hair cells, thus involved in hearing and balance. Nevertheless, Wu et al. [28] informed that the available results argue against the participation of any of the mouse TRP channels in hair cell transduction. Conversely, other studies suggest that TRPC3 and TRPC6 are required for the normal function of cells involved in hearing and are potential components of mechanotransducing complexes [29]. TRPC1, TRPC3, TRPC5, and TRPC6 channels contribute to auditory mechanosensation in a combinatorial manner but have no direct role in cochlear mechanotransduction [30].

Six members of this superfamily, TRPA1, TRPM8, TRPV1, TRPV2, TRPV3, and TRPV4, seem to participate in temperature sensing and TRPV1, TRPV2, TRPV3, and TRPV4 have incompletely overlapping functions over a broad thermal range from warm to hot [3, 7, 23, 31]. TRPA1 and TRPM8 respond to cool and cold, TRPV1 and TRPV2 are activated by painful levels of heat (>43°C and > 52°C, respectively), TRPV3 and TRPV4 respond to non-painful warmth (33–39°C), TRPM8 is activated by non-painful cool temperatures (<25°C), and TRPA 1 is activated by painful cold (<18°C; [32]). As a poikilotherm that lives in water, the detection of small fluctuations in the temperature can be of capital importance in zebrafish for survival.

Most of the authors suggest that chemosensation is determined primarily by the chemical activation of nociceptors and thermoreceptors and that activation by chemicals involves the direct activity of an ion channel by chemical stimuli: the so-called ionotropic transduction [33, 34]. TRP channels play integral roles in transducing chemical stimuli, giving rise to sensations of taste, irritation, warmth, coolness, and pungency. Among them are TRPM5, TRPV1, TRPA1, TRPM8, TRPV3, and TRPV4 [35, 36]. In this regard, TRPV1 is responsive to noxious stimuli and various chemical agents [31, 37–40].

3. The superfamily of acid-sensing ion channels (ASICs)

ASICs are voltage-insensitive, amiloride-sensitive Na^+-selective cation channels that monitor moderate deviations from the physiological values of extracellular pH [24, 41, 42]. In mammals six ASIC proteins encoded by four genes have been identified: ASIC1a, ASIC1b, ASIC2a, ASIC2b, ASIC3, and ASIC4 which differ in their kinetics, external pH sensitivity, tissue distribution, and pharmacological properties [43, 44]. The pH values required for half-maximal activation are 6.2–6.8 for ASIC1a, 5.9–6.2 for ASIC1b, 4.9 for ASIC2a, and 6.5–6.7 for ASIC3 [45, 46]. This capability of respond to minimal variation in pH might be of capital importance for survival in the aquatic environment. Structurally, ASICs consist of two transmembrane domains and a large extracellular loop [47], and in addition to their roles as detectors of pH variations, some ASICs may work as mechanosensors (or are required for mechanosensation) and nociceptors [47–53].

The members of these two families of ion channels that exhibit mechanosensitivity, thermosensitivity, and chemosensitivity are summarized in **Table 1**.

Mechanosensitivity	Thermosensitivity	Chemosensitivity
TRPA1, TRPC1, TRPC3, TRPC5, TRPC6, TRPV1, TRPV2, TRPV4, TRPM3, TRPM4, TRPM7, TRPM8, TRPP1, TRPP2	TRPA1, TRPM8, TRPV1, TRPV2, TRPV3, TRPV4	TRPM5, TRPV1, TRPA1, TRPM8, TRPV3, TRPV4
ASIC1, ASIC2, ASIC3	ASIC1, ASIC2, ASIC3	ASIC1, ASIC2, ASIC3

Table 1. Members of the transient-receptor potential (TRP) ion channel and acid-sensing ion channel (ASIC) superfamilies that exhibit mechanosensitivity, thermosensitivity, and chemosensitivity.

	Neuromast	Inner ear	Olfactory epithelium	Taste buds	Retina
trpc1			+		INL, GCL
trpc2a			+		
trpc2b			+		
trpc3					+
trpc5a					INL, GCL
trpc5b					INL
trpc7a			+		
trpm1a					INL
trpm4b	+		+		
trpm5/TRPM5			+	+	
TRPA1	+	+			
TRPV4	HC, MC	NEC	CON, nsc	SC	AC
ASIC1	HC				
ASIC2	HC, nerves		nsc (cilia)	SC, nerves	INL, IPL, GCL
ASIC4	HC, MC			SC	PhR, GCL

AC, amacrine cells; CON, ciliated olfactory neurons; FR, photoreceptors; GCL, ganglion cell layer; HC, hair cells; INL, inner nuclear layer; IPL, inner plexiform layer; MC, mantle cells; NEC, neuroepithelial cells; SC, sensory cells; nsc, non-sensory (nonolfactory) cells.

Table 2. Detection of transient-receptor potential (TRP) ion channels and acid-sensing ion channels (ASICs) in the sensory organs of zebrafish in larval state and adults.

4. TRP and ASICs in zebrafish

Expression of individual TRP ion channels has been observed in many tissues evidencing their roles as multifunctional cellular sensor proteins and functional analysis revealed their participation in all kinds of sensory detection: thermodetection, mechanodetection, chemodetection, nociception, and light perception.

In zebrafish, TRP channel genes belonging to the melastatin (TRPM [54–56]) vanilloid [57], canonical (TRPC [58, 59]) ankyrin [60, 61] and TRPN (also known as NOMPC [62]) have been detected. But based on their distribution, not all these ion channels are expressed in SO and therefore are involved in sensory detection.

Phylogenetic analysis has revealed 11 *trpm* genes and 12 *trpc* genes in the zebrafish genome: simple genes were identified for *trmp2*, *trmp3*, *trpm5*, *trmp6*, and *trmp7*, duplicate orthologs were found for *trpm1* (*trpm1a* and *trmp1b*), and quadruplicate orthologs were found for *trpm4* (*trpm4a*, *trpm4b1*, *trpm4b2*, *trpm4b3*) [56]. Regarding the TRPC family, simple genes were identified for *trpc1* and *trpc3* and duplicate for *trpc2* (*trpc2a* and *trpc2b*), *trpc4* (*trpc4a* and *trpc4b*), *trpc5* (*trpc5a* and *trpc5b*), *trpc6* (*trpc6a* and *trpc6b*), and *trpc7* (*trpc7a* and *trpc7b*) [58, 59, 63, 64].

In zebrafish the orthologs and paralogs of the six ASIC proteins detected in mammals have been identified and are denominated zASICs: ZASIC1.1, zASIC1.2, zASIC1.3, zASIC2, zASIC4.1, and zASIC4.2. The six proteins encoded by these genes have similar predicted molecular masses (~60 kDa) and share 60–75% amino acid sequence with rat and human ASICs [65].

5. TRP and ASICs in the sensory organs of zebrafish

Whole-mount in situ hybridization experiments as well as immunohistochemistry revealed the occurrence of TRPs and ASICs in SO of zebrafish as well as changes in the pattern of expression between developing and adult animals. Furthermore, functional analyses have demonstrated the involvement of some ion channels in different modalities of sensitivity.

5.1. Lateral line, superficial neuromasts, and ear

The mechanosensory cells in zebrafish are grouped into superficial and deep neuromasts that form the lateral line system (LLS [66, 67]) and in the sensory epithelia of the inner ear [68, 69]. The neuromasts and the neuroepithelium of the inner ear are filled with specific hair sensory cells that sense water flow and movement and endolymph movement, respectively. Thus, hair sensory cells are functionally mechanoreceptors where the conversion of mechanical stimuli into electrochemical signals, i.e., mechanotransduction, takes place, presumably because of the presence of mechanotransducer ion channels.

The role of TRPA1 in zebrafish mechanotransduction is still unclear. Using morpholino anti-sense oligonucleotides, it was observed that TRPA1 is required for inner ear and LLS hair cell function since these animals showed deafness [60]. In contrast, trpa1a;trpa1b doubly homozygous mutant zebrafish larvae have normal hair cell function [61]. The mechanosensory roles of TRPA1 in zebrafish are not supported by results obtained in TRPA1 knockout mice in which inner ear hair cell function was normal [70, 71]. A preliminary study of our group failed to demonstrate TRPM8 in sensory cells of the inner ear (**Figure 1e**).

TRPV4 is involved in mechanic and chemical sensation but also responds to warm temperatures and acidic pH (see [72]). Whole-mount in situ hybridization in zebrafish embryos showed expression of *trpv4* in neuromasts of LLS at 24 hours post fertilization (hpf), which decreases at 3 days post-fecunation (dpf) but remains at residual levels [57]. Using immunohistochemistry, TRPV4 was detected in neuromasts showing two patterns of distribution: in mantle cells alone or in a subset of hair sensory cells in addition to the mantle cells; the superficial neuromasts (pit organs) also displayed TRPV4 immunostaining in both mantle and hair cells [73] (**Figure 1a**). In the inner ear, TRPV4 immunoreactivity was observed in some hair cells of the macula and in a subpopulation of hair cells in the cristae ampullaris of the three semicircular canals [73].

Regarding ASICs specific immunoreactivity for ASIC1 and ASIC3 (**Figure 1c**) was detected in the hair cells of LLS, while ASIC2 was restricted to the nerves supplying neuromasts (**Figure 1b**). Moreover, supporting and mantle cells, i.e., the non-sensory cells of the neuromasts, also displayed ASIC4 [74]. In the inner ear sensory cells, ASIC1 and ASIC3, but not ASIC4, were found in neurosensory cells (**Figure 1f, h**, and **i**), while ASIC2 was only found in the nerves supplying them (**Figure 1g**).

It is possible that these ion channels could account for the transduction of mechanical stimuli as specialized hair cells in the LLS neuromasts are able to detect water movements and vibrations comparable to hair cells in the mammalian inner ear. Moreover the occurrence of TRPV4 in the ear neuroepithelia claims for an involvement of this ion channel in hearing and balance.

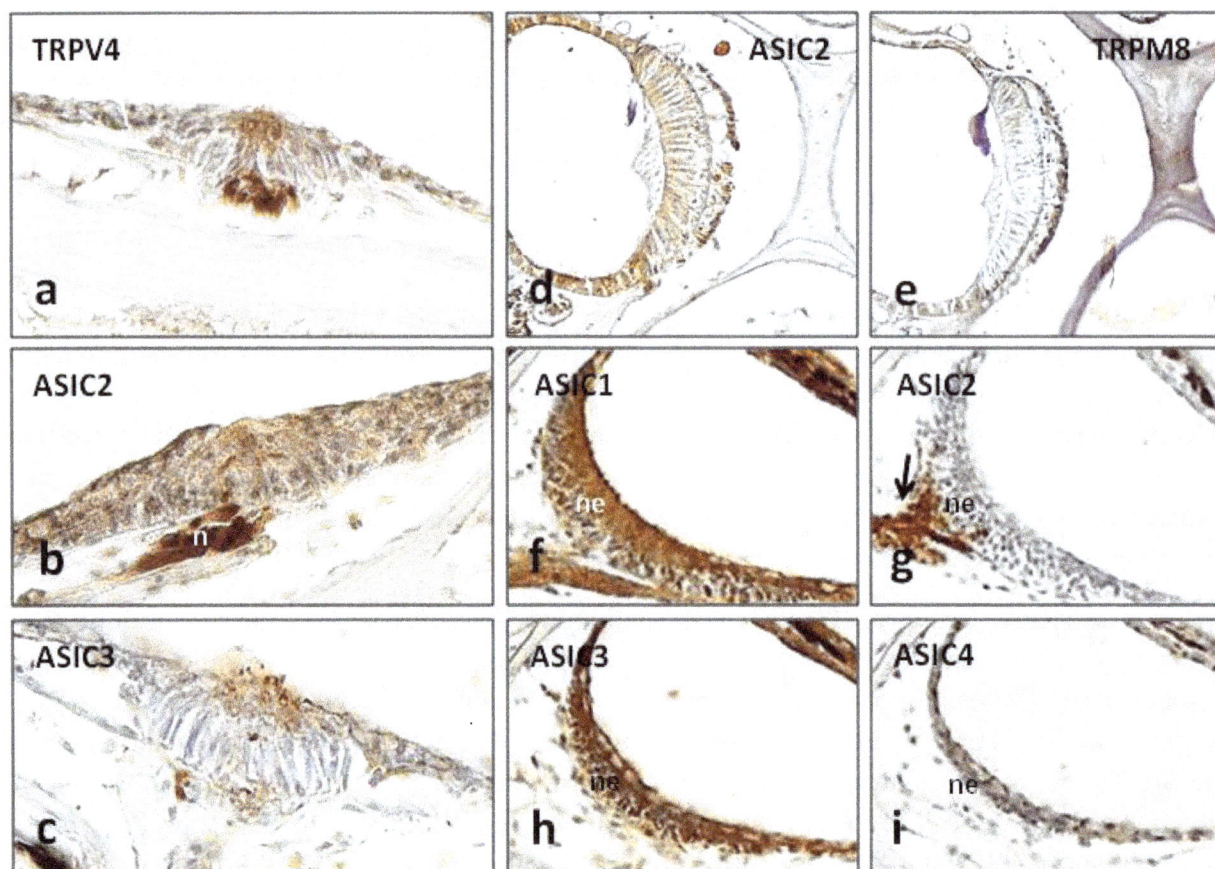

Figure 1. Immunohistochemical localization of ASICs and TRPM8 in the canal neuromasts (a–c) and inner ear neuroepithelia (ne) of adult zebrafish. N: nerve.

5.2. Olfactory epithelium and taste buds

The taste buds and the olfactory epithelium detect chemical changes in the aquatic environment. In zebrafish the chemosensory cells are grouped into taste buds [75] and in the lamellae of the olfactory epithelium [76]. Moreover, scattered solitary chemosensory cells are present in the skin [77].

The occurrence of *trpv*4 mRNA in the olfactory pit of zebrafish was observed during the embryonic period [57]. Thereafter, TRPV4 immunoreactivity was observed in ciliated olfactory neurons and in unidentified cells placed in the non-sensory olfactory epithelium but not the crypt neurons [73, 78] (**Figure 2b**). On the other hand, ASIC2 mRNA and protein were detected in the olfactory rosette of adult zebrafish. Specific ASIC2 hybridization was observed in the luminal pole of the non-sensory epithelium, especially in the cilia basal bodies, and immunoreactivity for ASIC2 was restricted to the cilia of the non-sensory cells; ASIC2 expression was always absent in the olfactory cells [79] (**Figure 2d**).

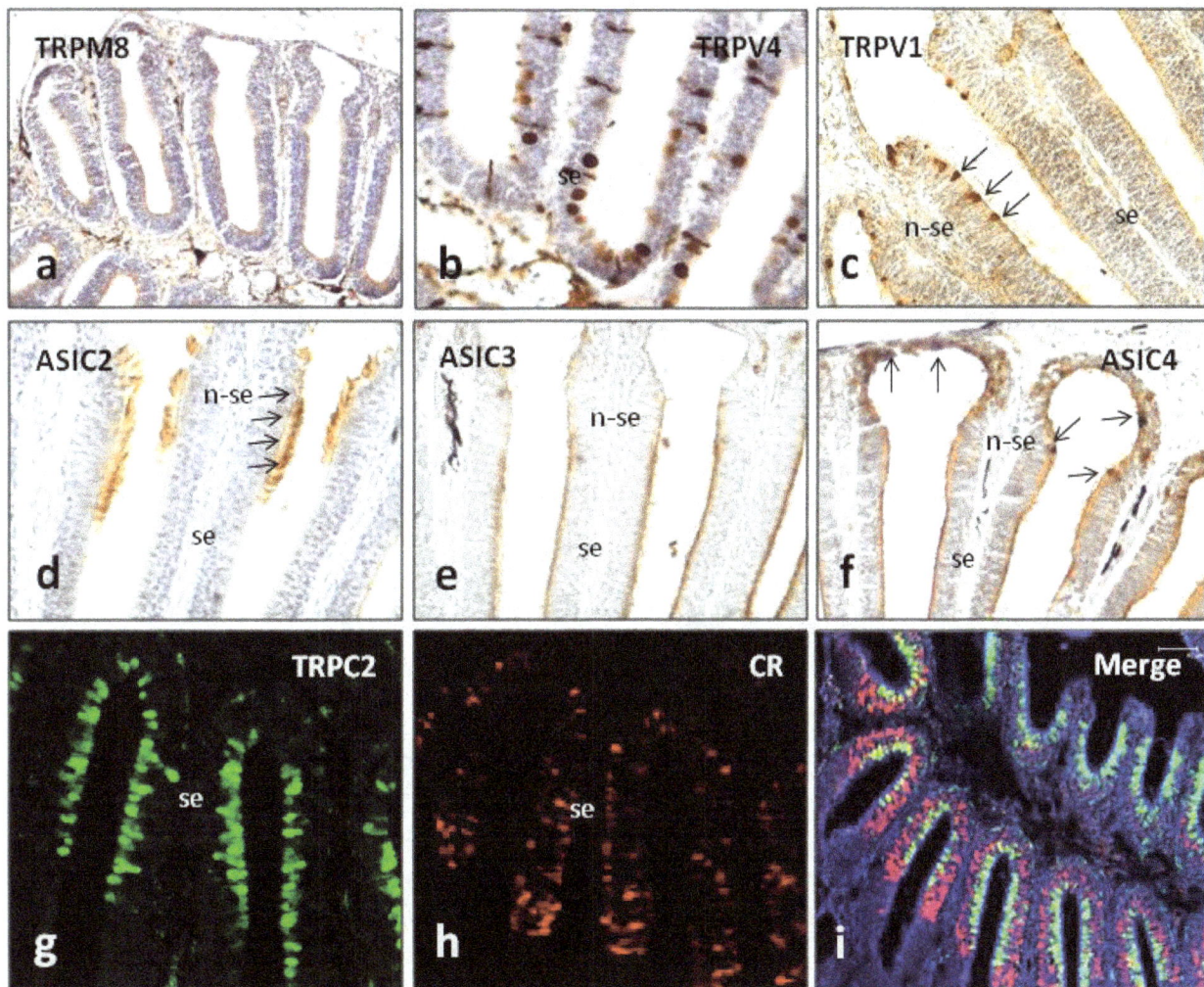

Figure 2. Immunohistochemical localization of TRP and ASICs in the olfactory epithelium of adult zebrafish. se, sensory epithelium; n-se, non-sensory epithelium.

The localization of TRPV4 in the olfactory epithelium suggests that it participates in the detection of chemical stimuli, including the odorant ones. Conversely the localization of ASIC2 suggests that it is not involved in olfaction. Since the cilia sense and transduce mechanical and chemical stimuli, ASIC2 expression in this location might be related to detection of aquatic environment, pH variations, or water movement through the nasal cavity.

Unpublished results for our laboratory also demonstrated the occurrence of TRPM8, ASIC3, and ASIC4 in the microvilli of the sensory epithelium and of TRPV1 in some unidentified cells of the non-sensory epithelium (**Figure 2a, c, e,** and **f**). Furthermore, we detected TRPC2 in a subpopulation of olfactory neurons different to the calretinin-positive ones (**Figure 2g–i**).

In fish taste receptor cells, different classes of ion channels have been detected which, like in mammals, presumably participate in the detection and/or transduction of chemical gustatory signals. The zebrafish homolog of TRPM5 (zfTRPM5) is expressed in cells of the taste buds [55]. TRPV4 has been also detected in the sensory cells of the cutaneous taste buds and in a subset of sensory cells in the oropharyngeal ones [73, 80] (**Figure 3a**). In addition TRPV4 was detected in the cutaneous solitary chemosensory cells [73]. Preliminary results of our group

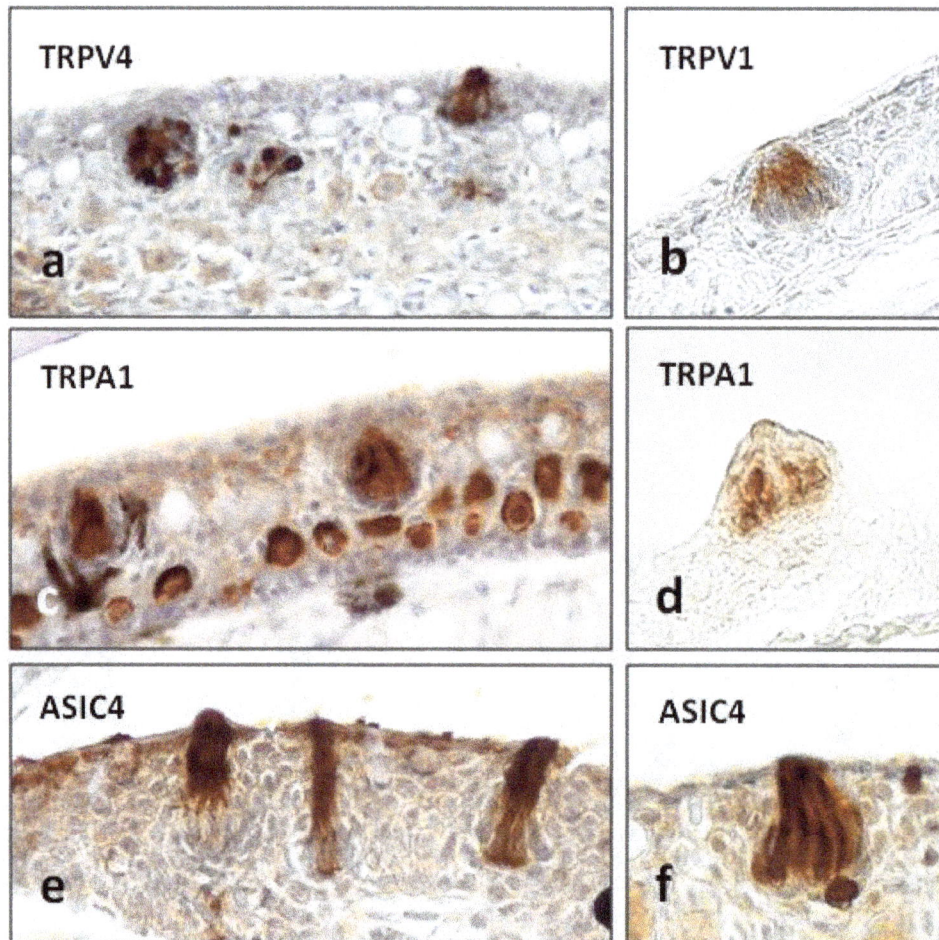

Figure 3. Immunohistochemical localization of TRP and ASICs in the taste buds of adult zebrafish.

have also detected TRPM1 in taste buds (**Figure 3b–d**). On the other hand, ASIC1 and ASIC3 were regularly absent from taste buds, whereas faint ASIC2 and robust ASIC4 immunoreactivities were detected in sensory cells (**Figure 3e** and **f**). Moreover, ASIC2 immunoreactivity was found in nerves supplying taste [81]. Since these ion channels are involved in the detection of sensory modalities other than olfaction, it can be hypothesized that taste cells sense stimuli other than those specific for taste.

5.3. Retina

Some TRP channels are present in the vertebrate retina. *trpC1* expression was observed in the ganglion cells as well as the inner nuclear layer of the eye, while *trpC6* was absent from SO [63]. Viña et al. [82] have investigated the expression and distribution of TRPV4 in the retina of zebrafish from 3 until 100 days post fertilization (dpf). Immunohistochemistry revealed the presence of TRPV4 in amacrine cells, localized in the inner nuclear layer and ganglion cell layers [73, 83] (**Figure 4a–c**). At 24 and 48 hpf, *trpm1a* was found expressed in different cells of the retina; thereafter at 3 dpf, it was expressed in the inner nuclear layer. On the other hand, *trpm1b* was initially expressed in cells of the outer retinal neuroepithelium and then in the inner nuclear layer [56].

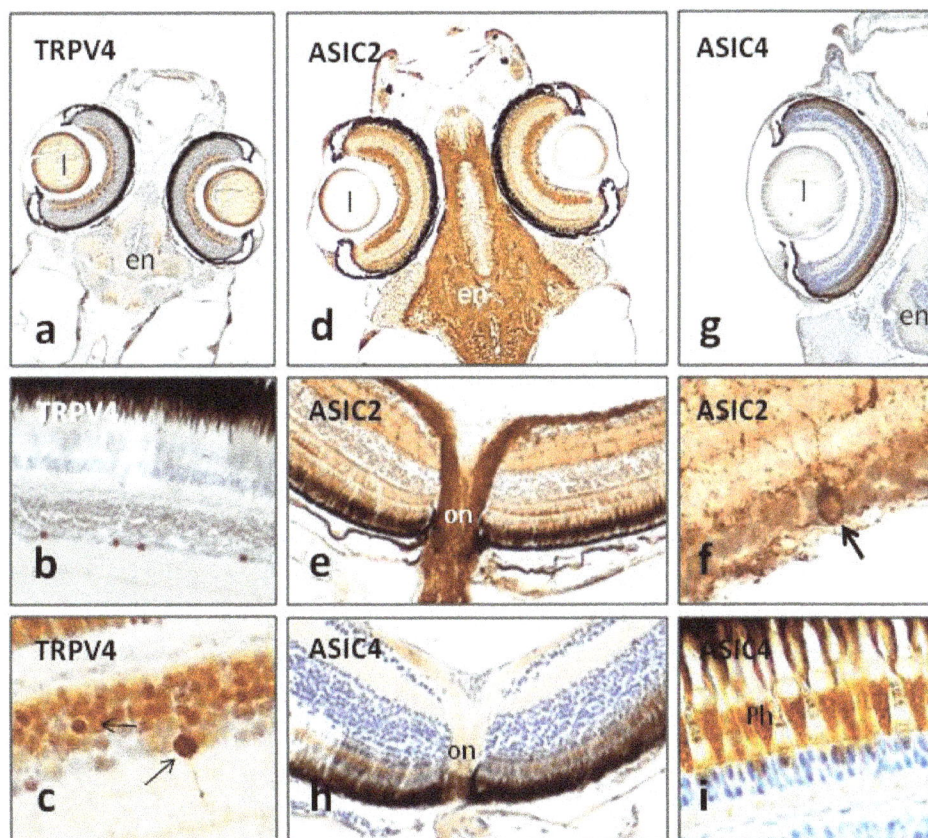

Figure 4. Immunohistochemical localization of TRP and ASICs in the retina of larvae (a, d, g) and adult zebrafish. en, encephalon; l, lens; on, optic nerve; Ph, photoreceptors.

Regarding ASICs, in the retina of zebrafish larvae, ASIC2 and ASIC4 were detected in the retinal ganglion cells [65]. *asic*1 mRNA and protein expressions were observed in the adult zebrafish retina using whole-mount *in situ* hybridization and immunohistochemistry study [84]. Viña et al. [82] in adult animals detected mRNA encoding ASIC2 and ASIC4.2 but not zASIC4.1. ASIC2 was found in the outer nuclear layer, the outer plexiform layer, the inner plexiform layer, the retinal ganglion cell layer, and the optic nerve. ASIC4 was expressed in the photoreceptor layer and to a lesser extent in the retinal ganglion cell layer (**Figure 4d–i**). Furthermore, the expression of both ASIC2 and ASIC4.2 was downregulated by light and darkness [82].

6. Concluding remarks

The sensory organs of zebrafish express multifunctional TRP and ASICs, most of them related to sensory modalities other than those expected for the sensory cells in which they are expressed. Based on the distribution of these multifunctional ion channels in SO, it seems they participate in multiple physiological functions as in mammals (mechanosensation, hearing, and temperature sensing) but furthermore have potential roles in olfaction, taste, and vision.

The ability to detect fluctuations in the aquatic temperature is critical to maintaining body temperature and avoiding injury in diverse animals from insects to mammals. In zebrafish ion channels are required for the sensation of heat [85]. Also of special importance for zebrafish is the detection of water movement. This function is classically attributed to the neuromast hair cells of the LLS. The occurrence of mechanosensitive ion channels in these cells supports this idea. But because the role of ASIC2 in mechanosensing and its presence in the cilia of the nonolfactory epithelium it is plausible suggest also a role of these cells in detecting water movement (see [79]). On the other hand, the presence of TRPV4 in sensory cells of the neuromasts and of the inner ear claims for an involvement of TRPV4 in mechanotransduction as suggested before [2, 72, 86] and is of particular importance since it potentially plays significant roles in human hearing [87].

Different TRP and ASICs have been observed in olfactory neurons and sensory taste cells of zebrafish, but its functions remain to be elucidated. However, in supporting a role of TRPV4 in olfaction in zebrafish, Ahmed et al. [88] and Nakashimo et al. [89] have detected TRPV1, TRPV2, TRPV3, and TRPV4 immunoreactivity in the olfactory epithelium of mice, and they suggest that TRPV channels may contribute to olfactory chemosensation. Of particular interest are the preliminary data presented here about the occurrence of TRPC2 in the olfactory epithelium of adult zebrafish. This ion channel is related to the detection of pheromones and sexual behavior in mammals [90, 91] and since zebrafish lack of a true vomeronasal organ age, sex, seasonal changes in TRPC2 must be analyzed in depth. Regarding the taste, and differently to mammals [35], the involvement of TRP ion channels has been poorly studied, while no great differences seem to exist with respect to ASICs (see [80]).

The expression of multiple TRP ion channels and ASICs in the developing and adult retina suggests they participate in vision. Fluctuations in pH play an important role in the retina, and for that reason, ASICs, and presumably also some TRPs, are thought to be involved in the fine-tuning of visual perception, adaptation to different light intensities, and phototransduction

[92]. As pH variations are also associated with pathological conditions, ASICs are likely to be involved in the pathogenesis of retinal diseases [93, 94], and its blockade may have a potential neuroprotective effect in ocular ischemic diseases [95].

Author details

Antonino Germanà[1], Juan D. Muriel[2], Ramón Cobo[3], Olivia García-Suárez[3], Juan Cobo[4,5] and José A. Vega[3,6]*

*Address all correspondence to: javega@uniovi.es

1 Dipartimento di Scienze Veterinarie, Sezione di Anatomia, Zebrafish Lab. Università di Messina, Italy

2 Sección de Diagnóstico por Imagen, Instituto Asturiano de Odontología, Oviedo, Spain

3 Departamento de Morfología y Biología Celular, Grupo SINPOS, Universidad de Oviedo, Spain

4 Instituto Asturiano de Odontología, Oviedo, Spain

5 Departamento de Cirugía y Especialidades Médico-Quirúrgicas, Universidad de Oviedo, Spain

6 Facultad de Ciencias de la Salud, Universidad Autónoma de Chile, Temuco, Chile

References

[1] Ostrander GK. The Laboratory Fish. London: Academic Press; 2000

[2] Damann N, Voets T, Nilius B. TRPs in our senses. Current Biology. 2008;**18**:R880-R889

[3] Nilius B, Szallasi A. Transient receptor potential channels as drug targets: From the science of basic research to the art of medicine. Pharmacological Reviews. 2014;**66**:676-814

[4] Jardín I, López JJ, Diez R, Sánchez-Collado J, Cantonero C, Albarrán L, Woodard GE, Redondo PC, Salido GM, Smani T, Rosado JA. TRPs in pain sensation. Frontiers in Physiology. 2017;**8**:392

[5] Liedtke WB. Chapter 22: TRPV channels' function in Osmo- and mechanotransduction. In: Liedtke WB, Heller S, editors. TRP Ion Channel Function in Sensory Transduction and Cellular Signaling Cascades. Boca Raton (FL): CRC Press; 2007

[6] Belmonte C, Viana F. Molecular and cellular limits to somatosensory specificity. Molecular Pain. 2008;**4**:14

[7] Wang H, Siemens J. TRP ion channels in thermosensation, thermoregulation and metabolism. Temperature (Austin). 2015;**2**:178-187

[8]   Morris AC. The genetics of ocular disorders: Insights from the zebrafish. Birth Defects Research. Part C, Embryo Today. 2011;**93**:215-228

[9]   Gestri G, Link BA, Neuhauss SC. The visual system of zebrafish and its use to model human ocular diseases. Developmental Neurobiology. 2012;**72**:302-327

[10] Richardson R, Tracey-White D, Webster A, Moosajee M. The zebrafish eye-a paradigm for investigating human ocular genetics. Eye (London, England). 2017;**31**:68-86

[11] Whitfield TT. Zebrafish as a model for hearing and deafness. Journal of Neurobiology. 2002;**53**:157-171

[12] Nicolson T. The genetics of hearing and balance in zebrafish. Annual Review of Genetics. 2005;**39**:9-22

[13] He Y, Bao B, Li H. Using zebrafish as a model to study the role of epigenetics in hearing loss. Expert Opinion on Drug Discovery. 2017;**12**:967-975

[14] Okada S. The taste system of small fish species. Bioscience, Biotechnology, and Biochemistry. 2015;**79**:1039-1043

[15] Orlando L. Odor detection in zebrafish. Trends in Neurosciences. 2001;**24**:257-258

[16] Cornell RA. Investigations of the in vivo requirements of transient receptor potential ion channels using frog and zebrafish model systems. Advances in Experimental Medicine and Biology. 2011;**704**:341-357

[17] Chen S, Chiu CN, McArthur KL, Fetcho JR, Prober DA. TRP channel mediated neuronal activation and ablation in freely behaving zebrafish. Nature Methods. 2016;**13**:147-150

[18] Jin P, Bulkley D, Guo Y, Zhang W, Guo Z, Huynh W, Wu S, Meltzer S, Cheng T, Jan LY, Jan YN, Cheng Y. Electron cryo-microscopy structure of the mechanotransduction channel NOMPC. Nature. 2017;**547**:118-122

[19] Li H. TRP channel classification. Advances in Experimental Medicine and Biology. 2017;**976**:1-8

[20] Clapham DE, Julius D, Montell C, Schultz G. International union of pharmacology. XLIX. Nomenclature and structure-function relationships of transient receptor potential channels. Pharmacological Reviews. 2005;**57**:427-450

[21] Hellmich UA, Gaudet R. Structural biology of TRP channels. Handbook of Experimental Pharmacology. 2014;**23**:963-990

[22] Eid SR, Cortright DN. Transient receptor potential channels on sensory nerves. Handbook of Experimental Pharmacology. 2009;**194**:261-281

[23] Nilius B, Owsianik G. The transient receptor potential family of ion channels. Genome Biology. 2011;**12**:218

[24] Lumpkin EA, Caterina MJ. Mechanisms of sensory transduction in the skin. Nature. 2007;**445**:858-865

[25] Arnadottir J, Chalfie M. Eukaryotic mechanosensitive channels. Annual Review of Biophysics. 2010;**39**:111-137

[26] Delmas P, Coste B. Mechano-gated ion channels in sensory systems. Cell. 2013;**155**:278-284

[27] Ranade SS, Syeda R, Patapoutian A. Mechanically activated ion channels. Neuron. 2015;**87**:1162-1179

[28] Wu X, Indzhykulian AA, Niksch PD, Webber RM, Garcia-Gonzalez M, Watnick T, Zhou J, Vollrath MA, Corey DP. Hair-cell mechanotransduction persists in TRP channel knockout mice. PLoS One. 2016;**11**:e0155577

[29] Quick K, Zhao J, Eijkelkamp N, Linley JE, Rugiero F, Cox JJ, Raouf R, Gringhuis M, Sexton JE, Abramowitz J, Taylor R, Forge A, Ashmore J, Kirkwood N, Kros CJ, Richardson GP, Freichel M, Flockerzi V, Birnbaumer L, Wood JN. TRPC3 and TRPC6 are essential for normal mechanotransduction in subsets of sensory neurons and cochlear hair cells. Open Biology. 2012;**2**:120068

[30] Sexton JE, Desmonds T, Quick K, Taylor R, Abramowitz J, Forge A, Kros CJ, Birnbaumer L, Wood JN. The contribution of TRPC1, TRPC3, TRPC5 and TRPC6 to touch and hearing. Neuroscience Letters. 2016;**610**:36-42

[31] Voets T. Quantifying and modeling the temperature-dependent gating of TRP channels. Reviews of Physiology, Biochemistry and Pharmacology. 2012;**162**:91-119

[32] Palkar R, Lippoldt EK, McKemy DD. The molecular and cellular basis of thermosensation in mammals. Current Opinion in Neurobiology. 2015;**34**:14-19

[33] Wood JN, Docherty R. Chemical activators of sensory neurons. Annual Review of Physiology. 1997;**59**:457-482

[34] Lee Y, Lee CH, Oh U. Painful channels in sensory neurons. Molecules and Cells. 2005;**20**:315-324

[35] Roper SD. TRPs in taste and chemesthesis. Handbook of Experimental Pharmacology. 2014;**223**:827-871

[36] Lehmann R, Schöbel N, Hatt H, van Thriel C. The involvement of TRP channels in sensory irritation: A mechanistic approach toward a better understanding of the biological effects of local irritants. Archives of Toxicology. 2016;**90**:1399-1413

[37] Reid G. ThermoTRP channels and cold sensing: What are they really up to? Pflügers Archiv. 2005;**451**:250-263

[38] Nieto-Posadas A, Jara-Oseguera A, Rosenbaum T. TRP channel gating physiology. Current Topics in Medicinal Chemistry. 2011;**11**:2131-2150

[39] Wetsel WC. Sensing hot and cold with TRP channels. International Journal of Hyperthermia. 2011;**27**:388-398

[40] Vay L, Gu C, McNaughton PA. The thermo-TRP ion channel family: Properties and therapeutic implications. British Journal of Pharmacology. 2012;**165**:787-801

[41] Lingueglia E. Acid-sensing ion channels in sensory perception. The Journal of Biological Chemistry. 2007;**282**:17325-17329

[42] Baron A, Lingueglia E. Pharmacology of acid-sensing ion channels—Physiological and therapeutical perspectives. Neuropharmacology. 2015;**94**:19-35

[43] Krishtal O. The ASICs: Signaling molecules? Modulators? Trends in Neurosciences. 2003;**26**:477-483

[44] Krishtal O. Receptor for protons: First observations on acid sensing ion channels. Neuropharmacology. 2015;**94**:4-8

[45] Kress M, Waldmann R. Acid sensing ionic channels. Current Topics in Membranes. 2006;**57**:241-276

[46] Hanukoglu I. ASIC and ENaC type sodium channels: Conformational states and the structures of the ion selectivity filters. The FEBS Journal. 2017;**284**:525-545

[47] Sherwood TW, Frey EN, Askwith CC. Structure and activity of the acid-sensing ion channels. American journal of physiology. Cell physiology. 2012;**303**:C699-C710

[48] Wemmie JA, Price MP, Welsh MJ. Acid-sensing ion channels: Advances, questions and therapeutic opportunities. Trends in Neurosciences. 2006;**29**:578-586

[49] Holzer P. Acid-sensitive ion channels and receptors. Handbook of Experimental Pharmacology. 2009;**194**:283-332

[50] Holzer P. Acid sensing by visceral afferent neurones. Acta Physiologica. 2011;**201**:63-75

[51] Zha XM. Acid-sensing ion channels: Trafficking and synaptic function. Molecular Brain. 2013;**6**(1)

[52] Holzer P, Izzo AA. The pharmacology of TRP channels. British Journal of Pharmacology. 2014;**171**:2469-2473

[53] Omerbašić D, Schuhmacher LN, Bernal Sierra YA, Smith ES, Lewin GR. ASICs and mammalian mechanoreceptor function. Neuropharmacology. 2015;**94**:80-86

[54] Elizondo MR, Arduini BL, Paulsen J, MacDonald EL, Sabel JL, Henion PD, Cornell RA, Parichy DM. Defective skeletogenesis with kidney stone formation in dwarf zebrafish mutant for trpm7. Current Biology. 2005;**15**:667-671

[55] Yoshida Y, Saitoh K, Aihara Y, Okada S, Misaka T, Abe K. Transient receptor potential channel M5 and phospholipaseC-beta2 colocalizing in zebrafish taste receptor cells. Neuroreport. 2007;**18**:1517-1520

[56] Kastenhuber E, Gesemann M, Mickoleit M, Neuhauss SC. Phylogenetic analysis and expression of zebrafish transient receptor potential melastatin family genes. Developmental Dynamics. 2013;**242**:1236-1249

[57] Mangos S, Liu Y, Drummond IA. Dynamic expression of the osmosensory channel trpv4 in multiple developing organs in zebrafish. Gene Expression Patterns. 2007;**7**(4):480

[58] Sato Y, Miyasaka N, Yoshihara Y. Mutually exclusive glomerular innervation by two distinct types of olfactory sensory neurons revealed in transgenic zebrafish. The Journal of Neuroscience. 2005;**25**:4889-4897

[59] Von Niederhäusern V, Kastenhuber E, Stäuble A, Gesemann M, Neuhauss SC. Phylogeny and expression of canonical transient receptor potential (TRPC) genes in developing zebrafish. Developmental Dynamics. 2013;**242**:1427-1441

[60] Corey DP, Garcia-Anoveros J, Holt JR, Kwan KY, Lin SY, Vollrath MA, Amalfitano A, Cheung EL, Derfler BH, Duggan A, Geleoc GS, Gray PA, Hoffman MP, Rehm HL, Tamasauskas D, Zhang DS. TRPA1 is a candidate for the mechanosensitive transduction channel of vertebrate hair cells. Nature. 2004;**432**:723-730

[61] Prober DA, Zimmerman S, Myers BR, BM MD Jr, Kim SH, Caron S, Rihel J, Solnica-Krezel L, Julius D, Hudspeth AJ, Schier AF. Zebrafish TRPA1 channels are required for chemosensation but not for thermosensation or mechanosensory hair cell function. The Journal of Neuroscience. 2008;**28**:10102-10110

[62] Sidi S, Friedrich RW, Nicolson T. NompC TRP channel required for vertebrate sensory hair cell mechanotransduction. Science. 2003;**301**:96-99

[63] Möller CC, Mangos S, Drummond IA, Reiser J. Expression of trpC1 and trpC6 orthologs in zebrafish. Gene Expression Patterns. 2008;**8**:291-296

[64] Petko JA, Kabbani N, Frey C, Woll M, Hickey K, Craig M, Canfield VA, Levenson R. Proteomic and functional analysis of NCS-1 binding proteins reveals novel signaling pathways required for inner ear development in zebrafish. BMC Neuroscience. 2009;**10**:27

[65] Paukert M, Sidi S, Russell C, Siba M, Wilson SW, Nicolson T, Gründer S. A family of acid-sensing ion channels from the zebrafish: Widespread expression in the central nervous system suggests a conserved role in neuronal communication. The Journal of Biological Chemistry. 2004;**279**:18783-18791

[66] Coombs S, New JG, Nelson MI. Information-processing demands in electrosensory and mechanosensory lateral line systems. Journal of Physiology, Paris. 2002;**96**:341-354

[67] Ghysen A, Dambly-Chaudière C. The lateral line microcosmos. Genes & Development. 2007;**21**:2118-2130

[68] Popper AN. Organization of the inner ear and auditory processing. In: Northcutt RG, Davis RE, editors. Fish Neurobiology. Brain Stem and Sense Organs. Vol. 1. Ann Arbor: University of Michigan Press; 1983. pp. 126-778

[69] Bang PI, Sewell WF, Malicki JJ. Morphology and cell type heterogeneities of the inner ear epithelia in adult and juvenile zebrafish (*Danio rerio*). The Journal of Comparative Neurology. 2001;**438**:173-190

[70] Bautista DM, Jordt SE, Nikai T, Tsuruda PR, Read AJ, Poblete J, Yamoah EN, Basbaum AI, Julius D. TRPA1 mediates the inflammatory actions of environmental irritants and proalgesic agents. Cell. 2006;**124**:1269-1282

[71] Kwan KY, Allchorne AJ, Vollrath MA, Christensen AP, Zhang DS, Woolf CJ, Corey DP. TRPA1 contributes to cold, mechanical, and chemical nociception but is not essential for hair-cell transduction. Neuron. 2006;**50**:277-289

[72] Plant TD, Strotmann R. TRPV4. Handbook of Experimental Pharmacology. 2007; **179**:189-205

[73] Amato V, Viña E, Calavia MG, Guerrera MC, Laurà R, Navarro M, De Carlos F, Cobo J, Germanà A, Vega JA. TRPV4 in the sensory organs of adult zebrafish. Microscopy Research and Technique. 2012;**75**:89-96

[74] Abbate F, Madrigrano M, Scopitteri T, Levanti M, Cobo JL, Germanà A, Vega JA, Laurà R. Acid-sensing ion channel immunoreactivities in the cephalic neuromasts of adult zebrafish. Annals of Anatomy. 2016;**207**:27-31

[75] Hansen A, Reutter K, Zeiske E. Taste bud development in the zebrafish, Danio rerio. Developmental Dynamics. 2002;**223**:483-496

[76] Hansen A, Zielinski BS. Diversity in the olfactory epithelium of bony fishes: Development, lamellar arrangement, sensory neuron cell types and transduction components. Journal of Neurocytology. 2005;**34**:83-208

[77] Kotrschal K, Krautgartner WD, Hansen A. Ontogeny of the solitary chemosensory cells in the zebrafish, *Danio rerio*. Chemical Senses. 1997;**22**:111-118

[78] Parisi V, Guerrera MC, Abbate F, Garcia-Suarez O, Viña E, Vega JA, Germanà A. Immunohistochemical characterization of the crypt neurons in the olfactory epithelium of adult zebrafish. Annals of Anatomy. 2014;**196**:178-182

[79] Viña E, Parisi V, Abbate F, Cabo R, Guerrera MC, Laurà R, Quirós LM, Pérez-Varela JC, Cobo T, Germanà A, Vega JA, García-Suárez O. Acid-sensing ion channel 2 (ASIC2) is selectively localized in the cilia of the non-sensory olfactory epithelium of adult zebrafish. Histochemistry and Cell Biology. 2015;**143**:59-68

[80] Levanti M, Randazzo B, Viña E, Montalbano G, Garcia-Suarez O, Germanà A, Vega JA, Abbate F. Acid-sensing ion channels and transient-receptor potential ion channels in zebrafish taste buds. Annals of Anatomy. 2016;**207**:32-37

[81] Viña E, Parisi V, Cabo R, Laurà R, López-Velasco S, López-Muñiz A, García-Suárez O, Germanà A, Vega JA. Acid-sensing ion channels (ASICs) in the taste buds of adult zebrafish. Neuroscience Letters. 2013;**536**:35-40

[82] Viña E, Parisi V, Sánchez-Ramos C, Cabo R, Guerrera MC, Quirós LM, Germanà A, Vega JA, García-Suárez O. Acid-sensing ion channels (ASICs) 2 and 4.2 are expressed in the retina of the adult zebrafish. Cell and Tissue Research. 2015b;**360**:223-231

[83] Sánchez-Ramos C, Guerrera MC, Bonnin-Arias C, Calavia MG, Laurà R, Germanà A, Vega JA. Expression of TRPV4 in the zebrafish retina during development. Microscopy Research and Technique. 2012;**75**:743-748

[84] Liu S, Wang MX, Mao CJ, Cheng XY, Wang CT, Huang J, Zhong ZM, Hu WD, Wang F, Hu LF, Wang H, Liu CF. Expression and functions of ASIC1 in the zebrafish retina. Biochemical and Biophysical Research Communications. 2014;**455**:353-357

[85] Gau P, Poon J, Ufret-Vincenty C, Snelson CD, Gordon SE, Raible DW, Dhaka A. The zebrafish ortholog of TRPV1 is required for heat-induced locomotion. The Journal of Neuroscience. 2013;**33**:5249-5260

[86] Orr AW, Helmke BP, Blackman BR, Schwartz MA. Mechanisms of mechanotransduction. Developmental Cell. 2006;**11**:20

[87] Cuajungco MP, Grimm C, Heller S. TRP channels as candidates for hearing and balance abnormalities in vertebrates. Biochimica et Biophysica Acta. 2007;**1772**:1022-1027

[88] Ahmed MK, Takumida M, Ishibashi T Hamamoto T, Hirakawa K. Expression of transient receptor potential vanilloid (TRPV) families 1, 2, 3 and 4 in the mouse olfactory epithelium. Rhinology. 2009;**47**:242-247

[89] Nakashimo Y, Takumida M, Fukuiri T, Anniko M, Hirakawa K. Expression of transient receptor potential channel vanilloid (TRPV) 1-4, melastin (TRPM) 5 and 8, and ankyrin (TRPA1) in the normal and methimazole-treated mouse olfactory epithelium. Acta Oto-Laryngologica. 2010;**130**:1278-1286

[90] Kiselyov K, van Rossum DB, Patterson RL. TRPC channels in pheromone sensing. Vitamins and Hormones. 2010;**83**:197-213

[91] Zufall F. TRPs in olfaction. Handbook of Experimental Pharmacology. 2014;**223**:917-933

[92] Ettaiche M, Guy N, Hofman P, Lazdunski M, Waldmann R. Acid-sensing ion channel 2 is important for retinal function and protects against light-induced retinal degeneration. The Journal of Neuroscience. 2004;**24**:1005-1012

[93] Tan J, Xu YP, Liu GP, Ye XH. Involvement of acid-sensing ion channel 1a in functions of cultured human retinal pigment epithelial cells. Journal of Huazhong University of Science and Technology. Medical Sciences. 2013;**33**:137-141

[94] Tan J, Ye X, Xu Y, Wang H, Sheng M, Wang F. Acid-sensing ion channel 1a is involved in retinal ganglion cell death induced by hypoxia. Molecular Vision. 2011;**17**:3300-3308

[95] Miyake T, Nishiwaki A, Yasukawa T, Ugawa S, Shimada S, Ogura Y. Possible implications of acid-sensing ion channels in ischemia-induced retinal injury in rats. Japanese Journal of Ophthalmology. 2013;**57**:120-125

Permissions

All chapters in this book were first published in RAZR, by InTech Open; hereby published with permission under the Creative Commons Attribution License or equivalent. Every chapter published in this book has been scrutinized by our experts. Their significance has been extensively debated. The topics covered herein carry significant findings which will fuel the growth of the discipline. They may even be implemented as practical applications or may be referred to as a beginning point for another development.

The contributors of this book come from diverse backgrounds, making this book a truly international effort. This book will bring forth new frontiers with its revolutionizing research information and detailed analysis of the nascent developments around the world.

We would like to thank all the contributing authors for lending their expertise to make the book truly unique. They have played a crucial role in the development of this book. Without their invaluable contributions this book wouldn't have been possible. They have made vital efforts to compile up to date information on the varied aspects of this subject to make this book a valuable addition to the collection of many professionals and students.

This book was conceptualized with the vision of imparting up-to-date information and advanced data in this field. To ensure the same, a matchless editorial board was set up. Every individual on the board went through rigorous rounds of assessment to prove their worth. After which they invested a large part of their time researching and compiling the most relevant data for our readers.

The editorial board has been involved in producing this book since its inception. They have spent rigorous hours researching and exploring the diverse topics which have resulted in the successful publishing of this book. They have passed on their knowledge of decades through this book. To expedite this challenging task, the publisher supported the team at every step. A small team of assistant editors was also appointed to further simplify the editing procedure and attain best results for the readers.

Apart from the editorial board, the designing team has also invested a significant amount of their time in understanding the subject and creating the most relevant covers. They scrutinized every image to scout for the most suitable representation of the subject and create an appropriate cover for the book.

The publishing team has been an ardent support to the editorial, designing and production team. Their endless efforts to recruit the best for this project, has resulted in the accomplishment of this book. They are a veteran in the field of academics and their pool of knowledge is as vast as their experience in printing. Their expertise and guidance has proved useful at every step. Their uncompromising quality standards have made this book an exceptional effort. Their encouragement from time to time has been an inspiration for everyone.

The publisher and the editorial board hope that this book will prove to be a valuable piece of knowledge for researchers, students, practitioners and scholars across the globe.

List of Contributors

Matti Vornanen, Jaakko Haverinen and Minna Hassinen
Department of Environmental and Biological Sciences, University of Eastern Finland, Finland

Zongming Ren
Institute of Environment and Ecology, Shandong Normal University, Jinan, PR China

Yuedan Liu
The Key Laboratory of Water and Air Pollution Control of Guangdong Province, South China Institute of Environmental Sciences, MEP, Guangzhou, PR China

Maria Violetta Brundo
Department of Biological, Geological and Environmental Science, University of Catania, Catania, Italy

Antonio Salvaggio
Experimental Zooprophylactic Institute of Sicily "A. Mirri", Catania, Italy

Decatur Foster and Kim Hanford Brown
Portland State University, Portland, OR, USA

Nikolay Popgeorgiev
Université de Lyon, Centre de recherche en cancérologie de Lyon, U1052 INSERM, UMR CNRS 5286, Université Lyon I, Centre Léon Bérard, Lyon, France

Benjamin Bonneau
Institut NeuroMyoGene, Université Claude Bernard Lyon 1, Centre National de la Recherche Scientifique, Unité Mixte de Recherche 5310, Institut National de la Santé et de la Recherche Médicale U1217, Lyon, France

Julien Prudent
Medical Research Council Mitochondrial Biology Unit, University of Cambridge, Wellcome Trust/MRC Building, Cambridge Biomedical Campus, Hills Road, Cambridge, United Kingdom

Germain Gillet
Université de Lyon, Centre de recherche en cancérologie de Lyon, U1052 INSERM, UMR CNRS 5286, Université Lyon I, Centre Léon Bérard, Lyon, France
Hospices civils de Lyon, Laboratoire d'anatomie et cytologie pathologiques, Centre Hospitalier Lyon Sud, chemin du Grand Revoyet, Pierre Bénite, France

Wan-Lun Taung and Jiann-Ruey Hong
Laboratory of Molecular Virology and Biotechnology, Institute of Biotechnology, National Cheng Kung University, Tainan City, Taiwan (R.O.C)
Department of Biotechnology and Bioindustry, National Cheng Kung University, Tainan City, Taiwan (R.O.C)

Jen-Leih Wu
Laboratory of Marine Molecular Biology and Biotechnology, Institute of Cellular and Organismic Biology, Taipei, Taiwan (R.O.C)

Dilan Celebi-Birand and Begun Erbaba
Interdisciplinary Graduate Program in Neuroscience, Aysel Sabuncu Brain Research Center, Bilkent University, Turkey
UNAM-National Nanotechnology Research Center and Institute of Materials Science and Nanotechnology, Bilkent University, Turkey
Department of Molecular Biology and Genetics Zebrafish Facility, Bilkent University, Turkey

Ahmet Tugrul Ozdemir
Department of Molecular Biology and Genetics Zebrafish Facility, Bilkent University, Turkey
Division of Cognitive Neurobiology, Center for Brain Research, Medical University of Vienna, Vienna, Austria

Hulusi Kafaligonul
Interdisciplinary Graduate Program in Neuroscience, Aysel Sabuncu Brain Research Center, Bilkent University, Turkey
Department of Molecular Biology and Genetics Zebrafish Facility, Bilkent University, Turkey
UMRAM-National Magnetic Resonance Research Center, Bilkent University, Turkey

Michelle Adams
Interdisciplinary Graduate Program in Neuroscience, Aysel Sabuncu Brain Research Center, Bilkent University, Turkey
UNAM-National Nanotechnology Research Center and Institute of Materials Science and Nanotechnology, Bilkent University, Turkey
Department of Molecular Biology and Genetics Zebrafish Facility, Bilkent University, Turkey
Department of Psychology, Bilkent University, Turkey

Conor Snyder, Reid Wilkinson, Amber Woodard, Andrew Lewis, Dallas Wood, Easton Haslam, Tyler Hogge, Nicolette Huntley, Jackson Pierce, Kayla Ranger, Luca Melendez, Townsend Wilburn, Brian Kiel, Ty Krug, Kaitlin Morrison, Aaliayh Lyttle, Wade E. Bell and James E. Turner
Department of Biology, Center for Molecular, Cellular, and Biological Chemistry, Virginia Military Institute, Lexington, VA, USA

Brandon Kar Meng Choo and Mohd. Farooq Shaikh
Neuropharmacology Research Laboratory, Jeffrey Cheah School of Medicine and Health Sciences, Monash University Malaysia, Bandar Sunway, Malaysia

Mayara Silveira, Jonatas Silveira, Thais Agues-Barbosa, Miguel Carvalho, Priscila Silva and Ana Luchiari
Physiology Department, Biosciences Center, Federal University of Rio Grande do Norte, Rio Grande do Norte, Brazil

Antonino Germanà
Dipartimento di Scienze Veterinarie, Sezione di Anatomia, Zebrafish Lab. Università di Messina, Italy

Juan D. Muriel
Sección de Diagnóstico por Imagen, Instituto Asturiano de Odontología, Oviedo, Spain

Ramón Cobo and Olivia García-Suárez
Departamento de Morfología y Biología Celular, Grupo SINPOS, Universidad de Oviedo, Spain

Juan Cobo
Instituto Asturiano de Odontología, Oviedo, Spain Departamento de Cirugía y Especialidades Médico-Quirúrgicas, Universidad de Oviedo, Spain

José A. Vega
Departamento de Morfología y Biología Celular, Grupo SINPOS, Universidad de Oviedo, Spain Facultad de Ciencias de la Salud, Universidad Autónoma de Chile, Temuco, Chile

Index